AF556732

APPLICATION OF
INVARIANT EMBEDDING
TO REACTOR PHYSICS

NUCLEAR SCIENCE AND TECHNOLOGY

A Series of Monographs and Textbooks

CONSULTING EDITOR

V. L. PARSEGIAN

Chair of Rensselaer Professor
Rensselaer Polytechnic Institute
Troy, New York

---

1. John F. Flagg (Ed.)
   CHEMICAL PROCESSING OF REACTOR FUELS, 1961
2. M. L. Yeater (Ed.)
   NEUTRON PHYSICS, 1962
3. Melville Clark, Jr., and Kent F. Hansen
   NUMERICAL METHODS OF REACTOR ANALYSIS, 1964
4. James W. Haffner
   RADIATION AND SHIELDING IN SPACE, 1967
5. Weston M. Stacey, Jr.
   SPACE-TIME NUCLEAR REACTOR KINETICS, 1969
6. Ronald R. Mohler and C. N. Shen
   OPTIMAL CONTROL OF NUCLEAR REACTORS, 1970
7. Ziya Akcasu, Gerald S. Lellouche, and Louis M. Shotkin
   MATHEMATICAL METHODS IN NUCLEAR REACTOR DYNAMICS, 1971
8. John Graham
   FAST REACTOR SAFETY, 1971
9. Akinao Shimizu and Katsutada Aoki
   APPLICATION OF INVARIANT EMBEDDING TO REACTOR PHYSICS, 1972

APPLICATION OF INVARIANT EMBEDDING TO REACTOR PHYSICS

AKINAO SHIMIZU AND KATSUTADA AOKI

Nippon Atomic Industry Group Co., Ltd.
NAIG Nuclear Research Laboratory
Kawasaki, Japan

1972

ACADEMIC PRESS • NEW YORK AND LONDON

ACADEMIC PRESS, INC.
111 Fifth Avenue, New York, New York 10003

United Kingdom Edition published by
ACADEMIC PRESS, INC. (LONDON) LTD.
24/28 Oval Road, London NW1

LIBRARY OF CONGRESS CATALOG CARD NUMBER: 72-187222

PRINTED IN THE UNITED STATES OF AMERICA

CONTENTS

CHAPTER THREE

SOLUTIONS OF EQUATIONS FOR SIMPLIFIED MODELS

CHAPTER FOUR

METHOD OF NUMERICAL SOLUTIONS

CHAPTER FIVE

REFLECTION AND TRANSMISSION OF GAMMA RAYS

PART B: *Application to Criticality Calculations*

CHAPTER SIX

INTRODUCTION

CHAPTER SEVEN

ONE-DIMENSIONAL PROBLEM

PREFACE

Recently, the systematic use of invariance concepts and functional equations in mathematical physics has been named "invariant embedding" by Richard Bellman. The invariant embedding approach is advantageous for computations by digital computor, since it reduces two-point boundary-value problems to initial-value problems.

The present monograph describes the application of the method of invariant embedding to radiation shielding and to criticality calculations of atomic reactors. The authors intend to show how this method has been applied to realistic problems, together with the results of applications which will be useful to shielding design.

The authors wish to express their sincere thanks to Professor Richard Bellman for suggesting that the present work be undertaken. They also express their thanks to Ryoich Wakabayashi, the director of NAIG Nuclear Research Laboratory for his constant support and encouragement. The authors owe the description of the design of a material testing reactor in Capter 6 to Mr. T. Omura of Nippon Atomic Industry Group Co., Ltd.

PART A *REFLECTION AND TRANSMISSION OF GAMMA RAYS BY SLABS*

CHAPTER ONE *INTRODUCTION*

1.1 Statement of the Problem

Suppose gamma rays from radioactive substances are incident on a shield. The reflection and transmission problem consists of determining the expected number of photons transmitted through or backscattered from the shield. The main problem is to determine reflection and transmission for a monoenergetic and monodirectional beam of incident photons, since any incident current can be represented as the sum of monoenergetic and monodirectional photons.

The photon energy of interest here lies in the range 0.01–10 MeV, which is typical for photons emitted from radioactive sustances or by fissions. Photons in this energy range interact with atoms in the medium by Compton scattering as well as by absorption reactions. Photons are decreased in energy and deflected in direction by Compton scattering. This is the root of all the difficulty in calculating transmission.

Figure 1.1 shows the absorption cross section of lead as a function of photon energy. It may be understood from the figure that the absorption cross section varies with energy so rapidly that an accurate treatment of the energy decrease due to scattering is necessary for transmission calculations.

The angular deviation caused by scattering also has a pronounced effect on the nature of gamma-ray penetration.

Figure 1.2 shows the differential cross section for Compton scattering as a function of the deflection angle. It is seen that the cross section is narrowly peaked in the forward direction at high energy.

When scattering is highly anisotropic, the direction of a photon after suffering many collisions still correlates to some extent with the direction before collisions. This relatively large but not perfect correlation in the photon direction requires an accurate treatment of the angular distribution at any point in the shield.

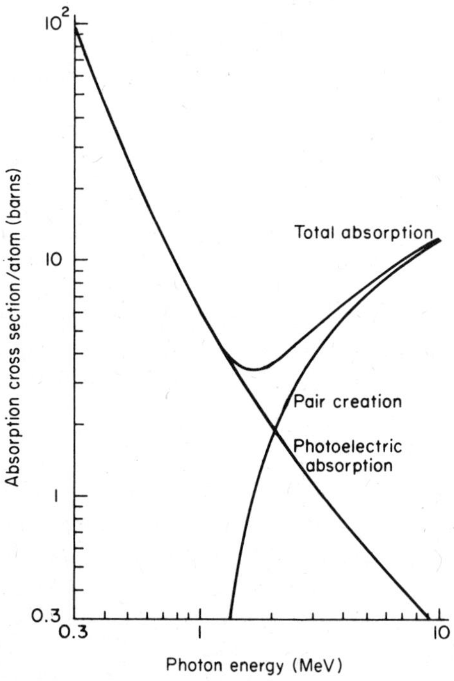

Fig. 1.1 *Absorption cross section for lead.*

The penetration problem of gamma rays is further complicated by the fact that there is a correlation between the deflection angle and the change of energy in the scattering.

Thus, the problem involves a function which depends on at least three variables, that is energy, an angular variable, and a spatial variable. These

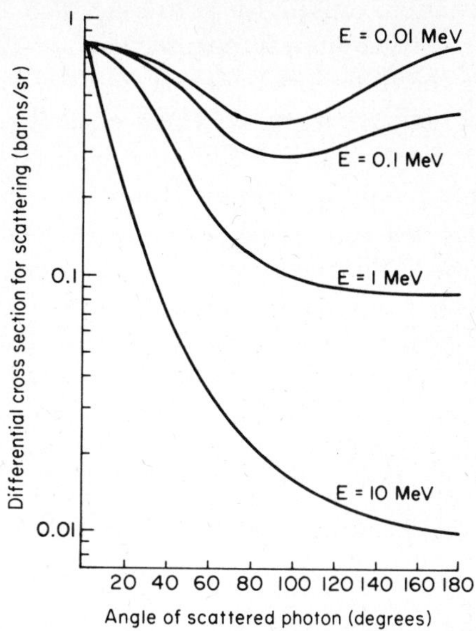

Fig. 1.2 *Differential cross section for Compton scattering.*

variables are usually not separable except for the special case. Accordingly, any accurate method of solving the problem must be numerical or semi-numerical in nature and involves extensive computations.

In subsequent chapters, we shall treat the reflection and transmission problem by the method of invariant embedding. The geometry of shields will be confined to slabs of finite thickness. The slab composition will be assumed either as homogeneous or heterogeneous. The method of invariant embedding may be applicable to shields of other one-dimensional geometries, yet realistic calculations made so far are limited to slabs.

1.2 The Interaction of Gamma Rays with Matter

Photons with energy from 10 keV to 10 MeV interact with matter by a number of processes. Among them, three processes are of primary importance to gamma-ray reflection and transmission. They are the photoelectric effect, pair production, and Compton scattering. The actual calculations of gamma-ray penetration have been made by considering only these three processes. These processes will be described briefly. More detailed descriptions will be found elsewhere [1–3].

1.2.1 The Photoelectric Effect

The photoelectric effect is a process of absorption of a photon by an atom. The energy of the photon is transferred to one of the electrons bounded to the atom. The cross section increases with the nuclear charge of the atom Z, roughly in proportion to Z^4–Z^5. As the photon energy rises, the cross section decreases rapidly from its maximum value at the binding energy. Figure 1.1 shows the photoelectric cross section for lead as a function of photon energy.

1.2.2 Pair Production

Pair production is a process in which a photon disappears and an electron–positron pair is created in the electric field of the nucleous or of the atomic electrons. The process has a threshold at 1.022 MeV, the sum of the rest energies of the electron and the positron, below which it cannot take place. The cross section for pair production rises rapidly as the photon energy increases above the threshold. Figure 1.1 shows the pair production cross section for lead as a function of energy. The cross section is also a rapidly rising function of the nuclear charge Z.

1.2.3 Compton Scattering

Compton scattering means the scattering of photons by free electrons. The cross section per atom is strictly proportional to Z.

The process is described more simply if the photon energy E is transformed into the photon wavelength λ given by the formula

$$\lambda \text{ (in Compton wavelength units)} = \frac{mc^2}{E} = \frac{0.5110}{E(\text{MeV})}. \tag{1.1}$$

Then the following relation between the deflection angle θ and the photon wavelengths before and after scattering can be derived from the conservation of energy and momentum,

$$\lambda - \lambda_0 = 1 - \cos\theta, \tag{1.2}$$

where λ is the wavelength after scattering and λ_0 the wavelength before scattering.

The cross section for scattering by a given angle, per unit solid angle, is given by the Klein–Nishina formula

$$\sigma(\theta, \lambda_0)\, d\Omega = \frac{3}{16\pi} \frac{\lambda_0^2}{\lambda^2} \left(\frac{\lambda_0}{\lambda} + \frac{\lambda}{\lambda_0} - \sin^2\theta \right). \tag{1.3}$$

The value of the cross section in Eq. (1.3) is given in Thomson units:

$$1 \text{ Thomson unit} = \frac{8\pi}{3} \left(\frac{e^2}{mc^2} \right) = 0.665 \quad \text{barn}. \tag{1.4}$$

The differential cross section $\sigma(\theta, \lambda_0)$ in Eq. (1.3) is plotted versus θ for various photon energies in Fig. 1.2.

In the transport calculations of photons for slabs, one needs the differential cross section for scattering from the photon direction specified by the cosine of the angle between the photon direction and the normal axis ω_0 to the direction ω and from the wavelength λ_0 to λ. This differential cross section is expressed as

$$\begin{aligned} &\sigma(\lambda_0, \omega_0 \to \lambda, \omega)\, d\lambda\, d\omega \\ &\quad = \int_0^{2\pi} d(\varphi - \varphi_0) \sigma(\theta, \lambda_0)\, \delta(\cos\theta - \lambda_0 + \lambda - 1)\, d\lambda\, d\omega, \end{aligned} \tag{1.5}$$

where $\delta(x)$ is the Dirac delta function, and φ_0 and φ are the azimuthal angles before and after scattering, respectively. The angle in Eq. (1.5) is a function of the variables ω, ω_0, and $(\varphi - \varphi_0)$ through the relation

$$\cos\theta = \omega\omega_0 + (1 - \omega_0^2)^{1/2}(1 - \omega^2)^{1/2} \cos(\varphi - \varphi_0). \tag{1.6}$$

By performing the integration in Eq. (1.5), we have

$$\begin{aligned} &\sigma(\lambda_0, \omega_0 \to \lambda, \omega) \\ &\quad = K(\lambda_0, \lambda) \cdot \mathrm{Re}[\pi(1 - \omega_0^2 - \omega^2 - \gamma^2 + 2\gamma\omega\omega_0)^{1/2}]^{-1}, \end{aligned} \tag{1.7}$$

where

$$K(\lambda_0, \lambda) = \frac{3}{8}\frac{\lambda_0^2}{\lambda^2}\left[\frac{\lambda_0}{\lambda} + \frac{\lambda}{\lambda_0} + 2(\lambda_0 - \lambda) + (\lambda_0 - \lambda)^2\right]$$

$$\text{for} \quad \lambda_0 \leq \lambda \leq \lambda_0 + 2,$$

$$= 0 \quad \text{otherwise,} \tag{1.8}$$

and

$$\gamma = 1 + \lambda_0 - \lambda. \tag{1.9}$$

The differential cross section $\sigma(\lambda_0, \omega_0 \to \lambda, \omega)$ is also expressed as

$$\sigma(\lambda_0, \omega_0 \to \lambda, \omega) = \tfrac{1}{2}K(\lambda_0, \lambda) \sum_{n=0}^{\infty} (2n + 1)P_n(\gamma)P_n(\omega_0)P_n(\omega), \tag{1.10}$$

where $P_n(x)$ is the Legendre polynomial of order n. Formula (1.10) is useful when the angular flux of photons is expanded by the Legendre polynomials. Expression (1.7), however, is more useful than Eq. (1.10) in calculating the differential cross section for given values of ω and ω_0. The series in Eq. (1.10) converges slowly, especially when γ is close to unity.

1.3 Classical Approach

In this section, we shall briefly describe successfully developed methods of solving gamma-ray reflection and transmission problems before applying the invariant embedding method. Stress will be placed on their advantages and limitations. More detailed descriptions of these methods will be found elsewhere [3, 4].

1.3.1 Method of Moments

The transport processes of gamma rays in a medium can be described by the well-known Boltzmann equation, which is the equation for the number flux of photons (the number of photons crossing a unit area perpendicular to the moving direction) at a phase point.

The reflection and transmission of gamma rays by a bounded medium can, in principle, be obtained by solving the Boltzmann equation subject to the boundary conditions. The boundary conditions for a slab are such that the flux of photons moving in the forward direction should be identical to the flux of incident photons at a surface and that the flux in the backward direction should vanish at the exit surface. This two-point boundary value problem is difficult to solve. In fact, the reflection coefficient of gamma rays has not yet been calculated successfully by solving the Boltzmann equation.

Accurate solutions of the Boltzmann equation have been obtained for the flux of photons from a plane or point source in an infinite homogeneous medium by using the method of moments proposed by Spencer and Fano [3, 5, 6]. Their formulation reduces the Boltzmann equation to a series of equations for functions involving only one variable. The series is terminated rigorously by the finite terms.

By using the moments method, an extensive series of calculations on gamma-ray penetration in infinite homogeneous media have been carried out by Goldstein and Wilkins [7]. Their results together with the boundary corrections obtained by Berger and Doggett [8] using the Monte Carlo method have furnished standard data for shielding designs.

The moments method, however, is applicable only to infinite homogeneous media. It breaks down completely for finite or heterogeneous media. The angular distribution of photons can be calculated by the moments method only with considerable difficulty.

1.3.2 Monte Carlo Method

The Monte Carlo method may be considered as a theoretical experiment. The reflection and transmission of gamma rays are calculated by tracing the history of a large number of photons in accordance with the probabilities given in terms of the cross sections for elementary processes. This method is quite feasible and applicable to a variety of problems. The theoretical procedures of the Monte Carlo method together with the results of calculations made so far are described in the article by Fano *et al.* [3].

The most serious restriction of the Monte Carlo method is that the fluctuation in the transmission obtained by this method is proportional to the inverse square root of the number of transmitted photons. Accordingly, the number of histories to be traced becomes exceedingly large for thick barriers.

Several refinements have been made to reduce the number of histories, yet it seems difficult to extend Monte Carlo calculations beyond a thickness of more than 10 mean free paths.

1.3.3 Other Methods

Peebles and Plesset [9] have made extensive calculations on the transmission of photons through iron and lead slabs by using the method of successive scattering. The transmissions of photons scattered in the medium up to four times are calculated by direct integrations. The contribution of photons scattered more than four times are estimated by extrapolations.

Peebles and Plesset also used an alternative procedure in which the transmission for thick slabs was calculated by successive operations of the reflection and transmission matrices for thin slabs. This method has recently been elaborated by Aronson and Yarmush [10] and also by Kataoka [11].

Aronson and Yarmush have been developing the transfer matrix method. They derive a formal expression for the transfer matrix for a slab of finite thickness in terms of the reflection and transmission matrices appropriate to an infinitesimally thin slab, which in turn are expressed directly in terms of the cross section for primary processes. Some numerical results were obtained by this method.

The guiding principle underlying their formulation seems essentially the same as that used in the invariant imbedding approach to be discussed later, although the formulations themselves are somewhat different from each other.

Kataoka calculated the reflection and transmission matrices for thin slabs by using the Monte Carlo method, and obtained the transmission through multiple layers by the matrix product.

The direct numerical integrations of the Boltzmann equations for photons has recently be attempted by Kataoka and Takeuchi [12] and Lathrop [13]. The main problem in this approach is to examine how many meshes in space, energy, and angle are required for solutions of adequate accuracy. The results of computations made by Kataoka and Takeuchi and also by Lathrop are encouraging. Further investigations are necessary to confirm the accuracies of numerical solutions at or near the boundaries.

1.4 Invariant Embedding Approach

The method of invariant embedding was introduced by the astrophysicist Ambarzumian [14] to solve the problem of diffuse reflection of light by a stellar atmosphere. His approach is radically new to the formulation of transport problems. He derived the integral equation for the reflection function of a semi-infinite plane-parallel atmosphere utilizing the principle of invariance, which means that the reflection function of a semi-infinite medium is invariant with respect to the addition of a layer of arbitrary thickness.

The principles of invariance were further generalized and formulated systematically into a general theory for radiative transfer by Chandrasekhar [15]. His theory was applied extensively to solve the problems of radiative transfer by himself and his co-workers.

Recently, the principles of invariances were further extended and generalized by Bellman and Kalaba [16, 17] and applied to the study of radiative transfer in a finite inhomogeneous atmosphere by themselves, Preisendorfer [18], and Ueno [19, 20].

The systematic use of invariance concepts and functional equations in mathematical physics is named "invariant embedding" by Bellman.

It has been emphasized by Bellman that the invariant embedding approach is advantageous for computations by digital computers, since it reduces two-point boundary value problems to initial value problems

An extensive series of calculations was made on the reflection of monoenergetic radiations by slabs of finite thickness in the case of isotropic scattering by Bellman *et al.* [21]. They used direct numerical integration to solve the equation instead of the semianalytical treatments developed by Chandrasekhar.

The method of invariant embedding originally introduced to solve the problem of light is applicable to problems of other radiations, such as gamma rays and neutrons, which have been studied extensively in association with radiology and nuclear engineering, in particular, reactor shielding design. The extension of the method from the problem of light to that of gamma rays or neutrons, however, is not straightforward. The problem of light is a one-velocity problem, since the wavelength of light changes on scattering by a negligible amount. The scattering law is nearly isotropic for light. On the other hand, gamma rays and neutrons change their energy appreciably by scattering. The variation of energy has a major effect on the penetration of gamma rays and neutrons. The scattering of gamma rays and neutrons is highly anisotropic at high energy. Furthermore, there is a correlation between the variation of energy and the deflection angle in scattering.

Applications of thc method of invariant embedding to the problems of neutrons and gamma rays with realistic energy and angle-dependent cross sections have been made recently. Shimizu and Mizuta [22–25] applied the method of invariant embedding to the reflection and transmission problem of gamma rays in slabs. An extensive series of calculations on the penetration of gamma rays through single- and double-layer slabs was made by this method [26, 27]. Mathews *et al.* [28] applied the method of invariant embedding to neutron shielding problems.

1.5 Differences between Boltzmann Approach and Invariant Embedding Approach

Before proceeding to the detailed description of the method of invariant embedding, we shall briefly discuss the differences between the Boltzmann approach and the invariant embedding approach.

Suppose we wish to compute the radiation intensity at the exit face of a slab due to an incident radiation. According to the Boltzmann approach, we compute the internal flux distribution in the slab by solving the Boltzmann equation subject to the boundary conditions which are given at both slab boundaries. The desired flux at the slab exit face cannot be obtained until we compute the flux at all points in the slab.

On the other hand, the invariant embedding approach is based on the equations for the reflection and transmission functions of the slab. The reflection and transmission functions are not functions of a spatial point in the slab like the flux, but are functions of the slab thickness. The equations for the functions are derived by considering how the functions vary if the slab thickness increases by an infinitesimal amount. The equations derived are the first-order differential equations with respect to slab thickness. The equation for the reflection function is nonlinear, but the equation for the transmission function is linear. The desired radiation intensity at the slab exit face can be obtained by integrating the equation for the transmission function from zero to the given thickness. It will be shown that this integration with respect to slab thickness can be accomplished quickly by utilizing the functional relation for the modified transmission function. The integration can be extended to a very large thickness without difficulty.

Advantages of the invariant embedding approach include:

(1) The numerical solution by computer requires less memory and time. In the Boltzmann approach, the flux at all points (practically at a finite number of points) in the slab should be computed simultaneously, whereas the transmission function for increasing thickness can be computed successively in the invariant embedding approach.

(2) An extension to a slab of large thickness (the deep penetration problem) is easy. As will be shown later, the transmission function for a thickness $2X$ can be obtained readily from the function for a thickness X by utilizing the functional relation. Thus, it is easy to extend the computation to a large thickness in the invariant embedding approach. On the other hand, it is difficult to extend the computation to a large thickness in the Boltzmann approach because of the requirement of excessive memory and time.

(3) The solutions are of a directly useful form. The reflection and transmission functions for single layers can be used directly to compute the functions for multilayer slabs. On the other hand, the solution for a single layer cannot be used directly to solve the multilayer problems in the Boltzmann approach.

Disadvantages of the invariant embedding approach include:

(1) It is not easy to apply in other than plane geometry, although the equation can be derived for spherical and cylindrical geometry. The equation for a two- or three-dimensional shield, such as a finite cylinder or cube, has not yet been derived.

(2) It does not directly yield the internal flux distribution in the slab, although it is possible to approximate it from the transmission functions.

CHAPTER TWO *FORMULATION OF THE PROBLEM BASED ON THE INVARIANT EMBEDDING PRINCIPLE*

2.1 Derivation of Basic Equations

2.1.1 Definition of Reflection and Transmission Functions

Suppose we have a homogeneous slab of thickness X which has an infinite extent in the y- and z-directions and is bounded in the x-direction by the planes $x = 0$ and $x = X$. Suppose, further, a photon is incident on the slab at the point $(0, y_0, z_0)$ on the surface $x = 0$ with energy E_0 in a direction specified by the cosine of the angle between the photon direction and the x-axis ω_0 and the azimuthal angle φ_0 (cf. Fig. 2.1).

The most precise solutions for the reflection and transmission of photons in the slab are given in terms of the reflection function

$$R(E, \omega, \varphi, y, z \mid E_0, \omega_0, \varphi_0, y_0, z_0; X)$$

(customarily called the differential albedo) and the transmission function

$$T(E, \omega, \varphi, y, z \mid E_0, \omega_0, \varphi_0, y_0, z_0; X).$$

These functions are defined so that

$$R(E, \omega, \varphi, y, z \mid E_0, \omega_0, \varphi_0, y_0, z_0; X)\, dE\, d\omega\, d\varphi\, dy\, dz \quad (2.1)$$

represents the expected number of photons emerging from the slab through the surface area $dy\,dz$ at the point $(0, y, z)$ on the surface $x = 0$ with energy E in the range dE in the backward direction (ω, φ)* within the solid angle $d\omega\,d\varphi$ due to the given incident photon, and that

$$T(E, \omega, \varphi, y, z \mid E_0, \omega_0, \varphi_0, y_0, z_0; X)\, dE\, d\omega\, d\varphi\, dy\, dz$$

represents the expected number of photons emerging from the slab through the surface $x = X$ in the forward direction.

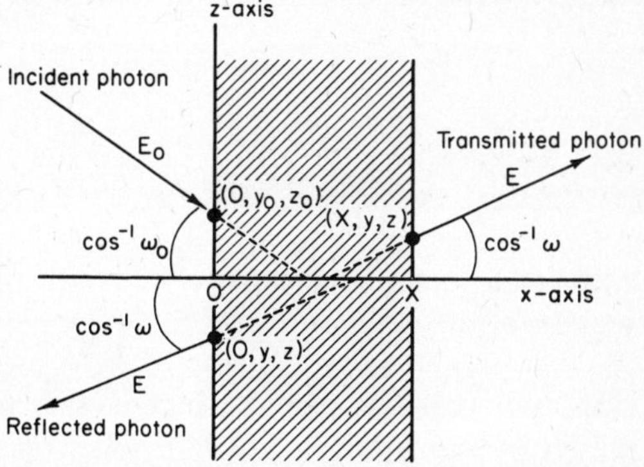

Fig. 2.1 *Reflection and transmission functions.*

Because of the symmetry of a homogeneous slab, the functions are expressed as

$$\begin{aligned} R(E, \omega, \varphi, y, z \mid E_0, \omega_0, \varphi_0, y_0, z_0; X) \\ = R(E, \omega \mid E_0, \omega_0; y - y_0, z - z_0, \varphi - \varphi_0; X), \end{aligned} \tag{2.2}$$

$$\begin{aligned} T(E, \omega, \varphi, y, z \mid E_0, \omega_0, \varphi_0, y_0, z_0; X) \\ = T(E, \omega \mid E_0, \omega_0; y - y_0, z - z_0, \varphi - \varphi_0; X). \end{aligned} \tag{2.3}$$

The reflection and transmission functions thus defined are too detailed to be obtained without excessive computation. We shall therefore confine ourselves to the calculation of the reflection and transmission functions integrated over the surface $x = 0$ or $x = X$ as well as over the relative

* Since the reflected photons are treated separately from the transmitted one, the variable ω is confined to the range $0 \leq \omega \leq 1$.

azimuth given by the equations

$$R(E, \omega \mid E_0, \omega_0; X) = \int_{-\infty}^{\infty} d(y - y_0) \int_{-\infty}^{\infty} d(z - z_0) \times \int_0^{2\pi} d(\varphi - \varphi_0) R(E, \omega \mid E_0, \omega_0; y - y_0, z - z_0, \varphi - \varphi_0; X), \tag{2.4}$$

$$T(E, \omega \mid E_0, \omega_0; X) = \int_{-\infty}^{\infty} d(y - y_0) \int_{-\infty}^{\infty} d(z - z_0) \times \int_0^{2\pi} d(\varphi - \varphi_0) T(E, \omega \mid E_0, \omega_0; y - y_0, z - z_0, \varphi - \varphi_0; X). \tag{2.5}$$

It should, however, be mentioned here that the equations for the functions

$$R(E, \omega, \varphi, y, z \mid E_0, \omega_0, \varphi_0, y_0, z_0; X)$$

and

$$T(E, \omega, \varphi, y, z \mid E_0, \omega_0, \varphi_0, y_0, z_0; X)$$

can also be derived by the invariant embedding method. Integration with respect to the variables $(y - y_0)$ and $(z - z_0)$ in Eqs. (2.4) and (2.5) may be interpreted as giving the number of photons emerging through all points on the surface due to an incident photon. Alternatively, the integral may be considered as representing the number of photons emerging per unit area due to a parallel beam of incident photons of the unit current density. Both interpretations are equivalent.

Similary, the functions integrated over the relative azimuth may be interpreted as representing the number of emergent photons going in all the directions within a given obliquity, or the number of emergent photons per unit angle due to axial symmetric incident photons from a conical source. The reflection and transmission functions introduced here are defined as the number current ratio, that is, the ratio of the number of photons crossing per unit area of the slab surface. The number flux ratio, which is defined as the ratio of the number of photons crossing per unit area perpendicular to the photon direction, is given by $(\omega_0/\omega) \times R(E, \omega \mid E_0, \omega_0; X)$ or $(\omega_0/\omega) T(E, \omega \mid E_0, \omega_0; X)$.

In parts of the following description, we shall use the condensed expressions $\mathbf{R}(X)$ and $\mathbf{T}(X)$ for the functions $R(E, \omega \mid E_0, \omega_0; X)$ and $T(E, \omega \mid E_0, \omega_0; X)$. The matrices $\mathbf{R}(X)$ and $\mathbf{T}(X)$ of an infinite order are subscripted by the continuous variables E and ω. A product of matrices in such expressions, that is $\mathbf{R}(X) \cdot \mathbf{T}(X)$, means a condensed expression of the function given by

$$\int_0^\infty dE' \int_0^1 d\omega' R(E, \omega \mid E', \omega'; X) T(E', \omega' \mid E_0, \omega_0; X). \tag{2.6}$$

2.1.2 Reflection and Transmission Functions for a Slab of Infinitesimal Thickness

We shall first calculate reflection and transmission functions for a slab of infinitesimal thickness dX. It will be shown that these functions can be expressed directly in terms of the basic cross section of the medium.

The functions can be expanded in a Taylor series of dX as

$$R(E, \omega \mid E_0, \omega_0; dX) = \dot{R}(E, \omega \mid E_0, \omega_0)\, dX + O(dX^2), \tag{2.7}$$

$$\begin{aligned} T(E, \omega \mid E_0, \omega_0; dX) = {} & \delta(E - E_0)\, \delta(\omega - \omega_0) - \dot{T}(E, \omega \mid E_0, \omega_0)\, dX \\ & + O(dX^2), \end{aligned} \tag{2.8}$$

where $\delta(X)$ is the Dirac delta function and $O(dX^2)$ means "of the order of dX^2." The functions $\dot{R}(E, \omega \mid E_0, \omega_0)$ and $\dot{T}(E, \omega \mid E_0, \omega_0)$ introduced in Eqs. (2.7) and (2.8) are formally defined by

$$\dot{R}(E, \omega \mid E_0, \omega_0) = \left[\frac{\partial}{\partial X} R(E, \omega \mid E_0, \omega_0; X)\right]_{X=0}, \tag{2.9}$$

$$\dot{T}(E, \omega \mid E_0, \omega_0) = -\left[\frac{\partial}{\partial X} T(E, \omega \mid E_0, \omega_0; X\right]_{X=0}. \tag{2.10}$$

These functions can be expressed in terms of the basic cross section as

$$\dot{R}(E, \omega \mid E_0, \omega_0) = \frac{1}{\omega_0} \Sigma_s(E_0, \omega_0 \to E, -\omega), \tag{2.11}$$

$$\begin{aligned} \dot{T}(E, \omega \mid E_0, \omega_0) = {} & \delta(E - E_0)\, \delta(\omega - \omega_0) \frac{\Sigma(E)}{\omega} \\ & - \frac{1}{\omega_0} \Sigma_s(E_0, \omega_0 \to E, \omega), \end{aligned} \tag{2.12}$$

where $\Sigma(E)$ is the macroscopic total cross section of the medium (the mass absorption coefficient) and $\Sigma_s(E_0, \omega_0 \to E, \omega)$ is the macroscopic differential cross section for scattering from energy E_0 to E and from the direction ω_0 to ω.

Equations (2.11) and (2.12) can be derived as follows: Suppose a parallel beam of monoenergetic photons with energy E_0 is incident on the slab in the direction ω_0. It is assumed that the flux density of the incident photons is unity. Then the flux of the transmitted photons is given by

$$\begin{aligned}\phi_t(E, \omega) = {} & \delta(E - E_0)\, \delta(\omega - \omega_0) e^{-\Sigma(E) dX/\omega} \\ & + \int_0^{dX} \frac{dx}{\omega}\, e^{-\Sigma(E)(dX - x)/\omega} \cdot \Sigma_s(E_0, \omega_0 \to E, \omega) e^{-\Sigma(E_0)x/\omega_0} \\ & + O(dX^2). \end{aligned} \tag{2.13}$$

The first term on the right-hand side of Eq. (2.13) represents unscattered photons and the second term singly scattered photons. The number of photons scattered more than once in a slab of thickness dX is of the order of dX^2. For sufficiently small thickness, Eq. (2.13) is reduced to

$$\begin{aligned}\phi_t(E, \omega) = {} & \delta(E - E_0)\, \delta(\omega - \omega_0)\left(1 - \frac{\Sigma(E)}{\omega}\, dX\right) \\ & + \frac{dX}{\omega}\, \Sigma_s(E_0, \omega_0 \to E, \omega) + O(dX^2). \end{aligned} \tag{2.14}$$

The flux of the reflected photons is likewise given by

$$\phi_r(E, \omega) = \frac{dX}{\omega}\, \Sigma_s(E_0, \omega_0 \to E, -\omega) + O(dX^2). \tag{2.15}$$

From Eqs. (2.14) and (2.15) and the definition of the reflection and transmission functions, we have

$$\begin{aligned}R(E, \omega \mid E_0, \omega_0; dX) & = \frac{\omega}{\omega_0}\, \phi_r(E, \omega) \\ & = \frac{dX}{\omega_0}\, \Sigma_s(E_0, \omega_0 \to E, -\omega) + O(dX^2), \end{aligned} \tag{2.16}$$

$$\begin{aligned}T(E, \omega \mid E_0, \omega_0; dX) = {} & \delta(E - E_0)\, \delta(\omega - \omega_0)\left(1 - \frac{\Sigma(E)}{\omega}\, dX\right) \\ & + \frac{dX}{\omega_0}\, \Sigma_s(E_0, \omega_0 \to E, \omega) + O(dX^2). \end{aligned} \tag{2.17}$$

By comparing Eqs. (2.7) and (2.8) with (2.16) and (2.17), we obtain the desired expressions.

2.1.3 Derivation of Equations for Reflection and Transmission Functions

We shall now derive equations for reflection and transmission functions. According to the invariant embedding principle, we shall consider how the functions vary if the thickness of a slab increases by an infinitesimal amount.

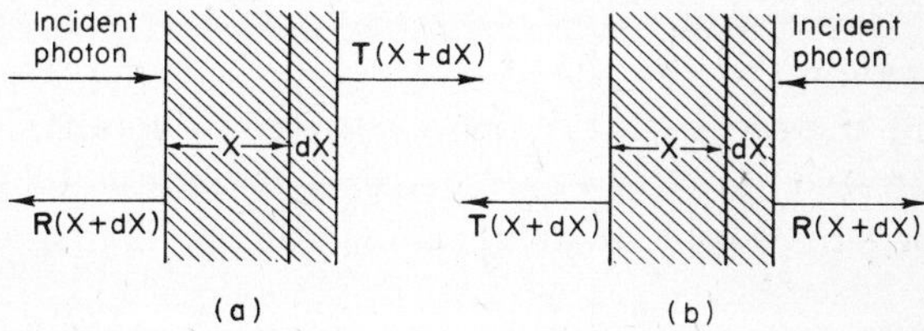

Fig. 2.2 *Reflection and transmission functions for a slab of thickness* $X + dX$.

Consider the case in which a parallel beam of photons is incident on a slab of thickness $(X + dX)$ as shown in Fig. 2.2a. The reflection function of the slab $\mathbf{R}(X + dX)$ is then expressed in terms of the reflection and transmission functions $\mathbf{R}(X)$, $\mathbf{T}(X)$, and $\mathbf{R}(dX)$ as

$$\mathbf{R}(X + dX) = \mathbf{R}(X) + \mathbf{T}(X)\mathbf{R}(dX)\mathbf{T}(X) + O(dX^2). \tag{2.18}$$

The first term on the right-hand side of Eq. (2.18) represents the photons reflected by a slab of thickness X, and the second term the photons first transmitted through a slab of thickness X and then reflected by a slab of thickness dX and then transmitted again through a slab of thickness X. The expected number of photons reflected more than once by a slab of thickness dX is of the order of dX^2. From Eqs. (2.18) and (2.7), we have

$$\frac{d}{dX}\mathbf{R}(X) = \mathbf{T}(X)\dot{\mathbf{R}}\mathbf{T}(X). \tag{2.19}$$

In a similar way, one can derive the equation for the transmission function,

$$\frac{d}{dX}\mathbf{T}(X) = [\mathbf{R}(X)\dot{\mathbf{R}} - \dot{\mathbf{T}}]\mathbf{T}(X). \tag{2.20}$$

Considering the case in which photons are incident on the slab through the opposite surface as shown in Fig. 2.2b, we obtain another set of equations,

$$\frac{d}{dX}\mathbf{R}(X) = \mathbf{R}(X)\dot{\mathbf{R}}\mathbf{R}(X) - \dot{\mathbf{T}}\mathbf{R}(X) - \mathbf{R}(X)\dot{\mathbf{T}} + \dot{\mathbf{R}}, \tag{2.21}$$

$$\frac{d}{dX}\mathbf{T}(X) = \mathbf{T}(X)[\dot{\mathbf{R}}\mathbf{R}(X) - \dot{\mathbf{T}}]. \tag{2.22}$$

Equation (2.21) is of particular importance, because it involves only the reflection function.

By inserting the expressions for the functions $\dot{\mathbf{R}}$ and $\dot{\mathbf{T}}$ given by Eqs. (2.11) and (2.12) into (2.19)–(2.22), we finally obtain the basic equations. They are

$$\begin{aligned}
&\frac{\partial}{\partial X} R(E, \omega \mid E_0, \omega_0; X) \\
&\quad = \int_0^\infty dE' \int_0^1 d\omega'\, T(E, \omega \mid E', \omega'; X) \\
&\qquad \times \int_0^\infty dE'' \int_0^1 \frac{d\omega''}{\omega''}\, \Sigma_s(E'', \omega'' \to E', -\omega')T(E'', \omega'' \mid E_0, \omega_0; X),
\end{aligned} \tag{2.23}$$

$$\begin{aligned}
&\frac{\partial}{\partial X} T(E, \omega \mid E_0, \omega_0; X) \\
&\quad = -\frac{\Sigma(E)}{\omega} T(E, \omega \mid E_0, \omega_0; X) \\
&\qquad + \int_0^\infty dE' \int_0^1 \frac{d\omega'}{\omega'}\, \Sigma_s(E', \omega' \to E, \omega)T(E', \omega' \mid E_0, \omega_0; X) \\
&\qquad + \int_0^\infty dE' \int_0^1 d\omega'\, R(E, \omega \mid E', \omega'; X) \\
&\qquad \times \int_0^\infty dE'' \int_0^1 \frac{d\omega''}{\omega''}\, \Sigma_s(E'', \omega'' \to E', -\omega')T(E'', \omega'' \mid E_0, \omega_0; X),
\end{aligned} \tag{2.24}$$

$$\begin{aligned}
&\frac{\partial}{\partial X} R(E, \omega \mid E_0, \omega_0; X) \\
&\quad = \frac{1}{\omega_0} \Sigma_s(E_0, \omega_0 \to E, -\omega) \\
&\qquad - \left[\frac{\Sigma(E)}{\omega} + \frac{\Sigma(E_0)}{\omega_0}\right] R(E, \omega \mid E_0, \omega_0; X) \\
&\qquad + \int_0^\infty dE' \int_0^1 \frac{d\omega'}{\omega'} \Sigma_s(E', \omega' \to E, \omega) R(E', \omega' \mid E_0, \omega_0; X) \\
&\qquad + \frac{1}{\omega_0} \int_0^\infty dE' \int_0^1 d\omega'\, R(E, \omega \mid E', \omega'; X) \Sigma_s(E_0, \omega_0 \to E', \omega') \\
&\qquad + \int_0^\infty dE' \int_0^1 d\omega'\, R(E, \omega \mid E', \omega'; X) \\
&\qquad \times \int_0^\infty dE'' \int_0^1 \frac{d\omega''}{\omega''} \Sigma_s(E'', \omega'' \to E', -\omega') R(E'', \omega'' \mid E_0, \omega_0; X),
\end{aligned} \tag{2.25}$$

$$\begin{aligned}
&\frac{\partial}{\partial X} T(E, \omega \mid E_0, \omega_0; X) \\
&\quad = -\frac{\Sigma(E_0)}{\omega_0} T(E, \omega \mid E_0, \omega_0; X) \\
&\qquad + \frac{1}{\omega_0} \int_0^\infty dE' \int_0^1 d\omega'\, T(E, \omega \mid E', \omega'; X) \Sigma_s(E_0, \omega_0 \to E', \omega') \\
&\qquad + \int_0^\infty dE' \int_0^1 d\omega'\, T(E, \omega \mid E', \omega'; X) \\
&\qquad \times \int_0^\infty dE'' \int_0^1 \frac{d\omega''}{\omega''} \Sigma_s(E'', \omega'' \to E', -\omega') R(E'', \omega'' \mid E_0, \omega_0; X),
\end{aligned} \tag{2.26}$$

The reflection and transmission functions evidently satisfy the initial conditions

$$R(E, \omega \mid E_0, \omega_0; 0) = 0, \tag{2.27}$$

$$T(E, \omega \mid E_0, \omega_0; 0) = \delta(E - E_0)\, \delta(\omega - \omega_0). \tag{2.28}$$

Equations of this type were first derived by Ambarzumian [14] and Chandrasekhar [15] for monoenergetic photons in a somewhat different way.

The fact that four equations are derived for two functions to be determined is due to the symmetry of a homogeneous slab.

2.1.4 Equations for an Inhomogeneous Slab

We shall derive equations for the functions of an inhomogeneous slab, whose composition is assumed to change only in the x-direction (the direction perpendicular to the surface of the slab).

Let the reflection and transmission functions of the slab viewed from the left-hand side (the functions defined for photons incident on the slab through the left surface) be $\mathbf{R}^+(X)$ and $\mathbf{T}^+(X)$ and those viewed from the right-hand side be $\mathbf{R}^-(X)$ and $\mathbf{T}^-(X)$. The equations for these four functions can be derived in ways similar to Eqs. (2.19) and (2.20). They are

$$\frac{d}{dX}\mathbf{R}^+(X) = T^-(X)\dot{\mathbf{R}}(X)\mathbf{T}^+(X), \tag{2.29}$$

$$\frac{d}{dX}\mathbf{T}^+(X) = [\mathbf{R}^-(X)\dot{\mathbf{R}}(X) - \dot{\mathbf{T}}(X)]\mathbf{T}^+(X), \tag{2.30}$$

$$\frac{d}{dX}\mathbf{R}^-(X) = \mathbf{R}^-(X)\dot{\mathbf{R}}(X)\mathbf{R}^-(X) - \dot{\mathbf{T}}(X)\mathbf{R}^-(X) - \mathbf{R}^-(X)\dot{\mathbf{T}}(X) + \dot{\mathbf{R}}(X), \tag{2.31}$$

$$\frac{d}{dX}\mathbf{T}^-(X) = \mathbf{T}^-(X)[\dot{\mathbf{R}}(X)\mathbf{R}^-(X) - \dot{\mathbf{T}}(X)], \tag{2.32}$$

where the functions $\dot{\mathbf{R}}(X)$ and $\dot{\mathbf{T}}(X)$ are given by

$$\dot{R}(E, \omega \mid E_0, \omega_0; X) = \frac{1}{\omega_0}\Sigma_s(E_0, \omega_0 \to E, -\omega; X), \tag{2.33}$$

$$\dot{T}(E, \omega \mid E_0, \omega_0; X) = \delta(E - E_0)\,\delta(\omega - \omega_0)\frac{\Sigma(E; X)}{\omega} - \frac{1}{\omega_0}\Sigma_s(E_0, \omega_0 \to E, \omega; X). \tag{2.34}$$

The variable X in the expressions for the macroscopic cross sections in Eqs. (2.33) and (2.34) indicates that the cross sections are for the composition at thickness X.

For a homogeneous slab, we have

$$\mathbf{R}^+(X) = \mathbf{R}^-(X),$$
$$\mathbf{T}^+(X) = \mathbf{T}^-(X).$$

Equations (2.29)–(2.32), then, are reduced to (2.19)–(2.22).

2.2 Reflection Function for a Semi-Infinite Medium and the Modified Transmission Functions

The direct numerical integrations of Eqs. (2.23)–(2.26) still require considerable computation, even though the integrations may be easier than the direct numerical integrations of the Boltzmann equation. In the present section, it will be shown that the equations can be reduced to more tractable forms.

2.2.1 Reflection Function for a Semi-Infinite Medium

The reflection function of a homogeneous slab will approach an asymptotic value as the thickness increases. The asymptotic value which is identical to the reflection function for a semi-infinite medium

$$R(E, \omega \mid E_0, \omega_0; \infty)$$

can be obtained directly by solving the integral equation

$$\begin{aligned}
&\left[\frac{\Sigma(E)}{\omega} + \frac{\Sigma(E_0)}{\omega_0}\right] R(E, \omega \mid E_0, \omega_0; \infty) \\
&\quad = \frac{1}{\omega_0} \Sigma_s(E_0, \omega_0 \to E, -\omega) \\
&\qquad + \int_0^\infty dE' \int_0^1 \frac{d\omega'}{\omega'} \Sigma_s(E', \omega' \to E, \omega) R(E', \omega' \mid E_0, \omega_0; \infty) \\
&\qquad + \frac{1}{\omega_0} \int_0^\infty dE' \int_0^1 d\omega' \, R(E, \omega \mid E', \omega'; \infty) \Sigma_s(E_0, \omega_0 \to E', \omega') \\
&\qquad + \int_0^\infty dE' \int_0^1 d\omega' \, R(E, \omega \mid E', \omega'; \infty) \\
&\qquad \times \int_0^\infty dE'' \int_0^1 \frac{d\omega''}{\omega''} \Sigma_s(E'', \omega'' \to E', -\omega') \, R(E'', \omega'' \mid E_0, \omega_0; \infty).
\end{aligned} \tag{2.35}$$

Equation (2.35) is derived from (2.25) based on the fact that the differential term $(d/dX)\mathbf{R}(X)$ vanishes when the thickness becomes infinite, since the reflection function for a semi-infinite medium is invariant with respect to the addition to the boundary of a layer of the same composition. This type of equation was first derived by Ambarzumian.

The reflection function for gamma rays approaches the asymptotic value rapidly as the slab thickness increases. Berger and Doggett [8] obtained the dependence of the reflection coefficient of gamma rays on slab thickness by using the Monte Carlo method. They found that the reflection coefficient of a slab of thickness greater than 1 or 2 mean free paths at source energy is practically the same as the reflection coefficient of a semi-infinite medium. This fact is confirmed experimentally by Bulatov and Garusov [29] and Fujita *et al.* [30]. Thus, the approximation

$$\mathbf{R}(X) \simeq \mathbf{R}(\infty) \tag{2.36}$$

may be applied to a slab of thickness X greater than 1 or 2 mean free paths.

2.2.2 Modified Transmission Function

It is found convenient in solving the transmission problem to introduce a function called the modified transmission function [23]. The modified transmission function $\tilde{\mathbf{T}}(E, \omega \mid E_0, \omega_0; X)$ is defined mathematically as the solution of the equation

$$\begin{aligned}\frac{\partial}{\partial X} \tilde{T}(E, \omega \mid E_0, \omega_0; X) &= -\frac{\Sigma(E)}{\omega} \tilde{T}(E, \omega \mid E_0, \omega_0; X) \\ &\quad + \int_0^\infty dE' \int_0^1 d\omega' \, C(E, \omega \mid E', \omega') \tilde{T}(E', \omega' \mid E_0, \omega_0; X)\end{aligned} \tag{2.37}$$

with the initial condition

$$\tilde{T}(E, \omega \mid E_0, \omega_0; 0) = \delta(E - E_0)\, \delta(\omega - \omega_0), \tag{2.38}$$

where the function $C(E, \omega \mid E_0, \omega_0)$ is given by

$$\begin{aligned}C(E, \omega \mid E_0, \omega_0) &= \frac{1}{\omega_0} \Sigma_s(E_0, \omega_0 \rightarrow E, \omega) \\ &\quad + \frac{1}{\omega_0} \int_0^\infty dE' \int_0^1 d\omega' \, R(E, \omega \mid E', \omega'; \infty) \Sigma_s(E_0, \omega_0 \rightarrow E', -\omega').\end{aligned} \tag{2.39}$$

Equation (2.37) has the same form as the equation for the transmission function (2.24) except that the reflection function $R(E, \omega \mid E_0, \omega_0; X)$ on the right-hand side of Eq. (2.24) is replaced by the reflection function for a semi-infinite medium $R(E, \omega \mid E_0, \omega_0; \infty)$.

The physical meaning of the modified transmission function will be given in the following.

Suppose there is a plane source which emits a photon per unit time per unit area with energy E_0 in the forward direction ω_0 ($\omega_0 \geq 0$) behind a homogeneous slab of thickness X. Suppose, further, a semi-infinite medium of the same composition as the slab is also placed behind the source (cf. Fig. 2.3). Let the expected number of photons that will emerge from the front surface of the slab per unit time per unit area with energy E in the direction ω be $T(E, \omega \mid E_0, \omega_0; X)$. Then it can be shown that this quantity satisfies Eq. (2.37) and the initial condition (2.38). Thus the quantity is the modified transmission function itself.

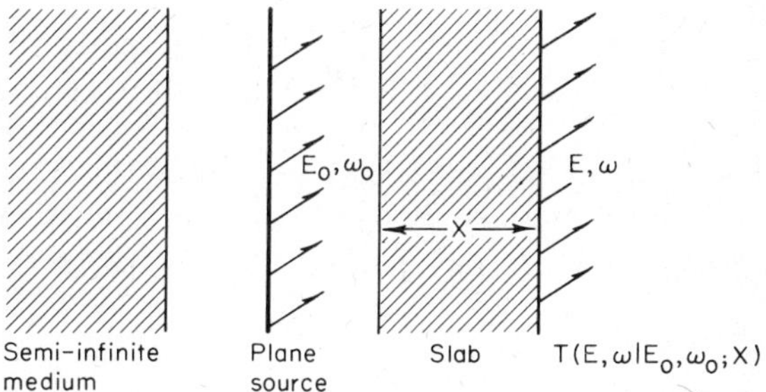

Fig. 2.3 *Modified transmission function.*

Equation (2.37) can be derived by considering how the intensity of the transmitted photons varies when a layer of infinitesimal thickness of the same composition is added to the front surface. The first term on the right-hand side of Eq. (2.37) represents the number of photons removed by the collision in the layer. The second term represents the number of photons reemerging in the forward direction after being scattered in the layer. The scattered photons consist of those scattered directly in the forward direction [the first term on the right-hand side of Eq. (2.39)] and of those scattered first in the backward direction and then reflected into the forward direction by the medium behind the front surface [the second term in Eq. (2.39)]. The reflection function of the medium behind the front surface is invariably given by $R(E, \omega \mid E_0, \omega_0; \infty)$ due to the semi-infinite medium behind the slab.

It should be emphasized here that the modified transmission function is defined as the expected number of transmitted photons per photon from the source. In the arrangement as given in Fig. 2.3, the intensity of photons incident on the slab is greater than that of the photons emitted directly from the source. The modified transmission function, therefore, is greater than the ordinary transmission function which is defined as the expected number of transmitted photons per incident photon.

From Eqs. (2.37) and (2.38), one can derive the important functional relation

$$\tilde{T}(E, \omega \mid E_0, \omega_0; X + X') = \int_0^\infty dE' \int_0^1 d\omega' \, \tilde{T}(E, \omega \mid E', \omega'; X') \tilde{T}(E', \omega' \mid E_0, \omega_0; X), \quad (2.40)$$

or using the condensed representation

$$\tilde{\mathbf{T}}(X + X') = \tilde{\mathbf{T}}(X')\tilde{\mathbf{T}}(X). \quad (2.41)$$

It should be emphasized here that the functional relation for the modified transmission function [Eq. (2.41)] has a simpler form than the corresponding equation for the ordinary transmission function and is expressed as

$$\mathbf{T}(X + X') = \mathbf{T}(X')[\mathbf{E} - \mathbf{R}(X)\mathbf{R}(X')]^{-1}\mathbf{T}(X),$$

where $\mathbf{E}$ is the unit matrix and $[\mathbf{E} - \mathbf{R}(X)\mathbf{R}(X')]^{-1}$ is the inverse of the matrix $[\mathbf{E} - \mathbf{R}(X)\mathbf{R}(X')]$. The simplicity of Eq. (2.41) is due to the fact that the modified transmission function is invariant with respect to the addition of a layer of the same composition between the semi-infinite medium and the plane source. This fact is illustrated schematically in Fig. 2.4, which also shows why the ordinary transmission function does not satisfy the functional relation (2.41).

The functional relation [Eq. (2.41)] is very useful in obtaining numerical solutions. By successive applications of Eq. (2.41), we obtain $\tilde{\mathbf{T}}(X)$, $\tilde{\mathbf{T}}(2X)$, $\tilde{\mathbf{T}}(4X)$, ..., $\tilde{\mathbf{T}}(2^nX)$. Thus, even if one starts from the function for a very thin slab, one can readily obtain the function for thick slabs.

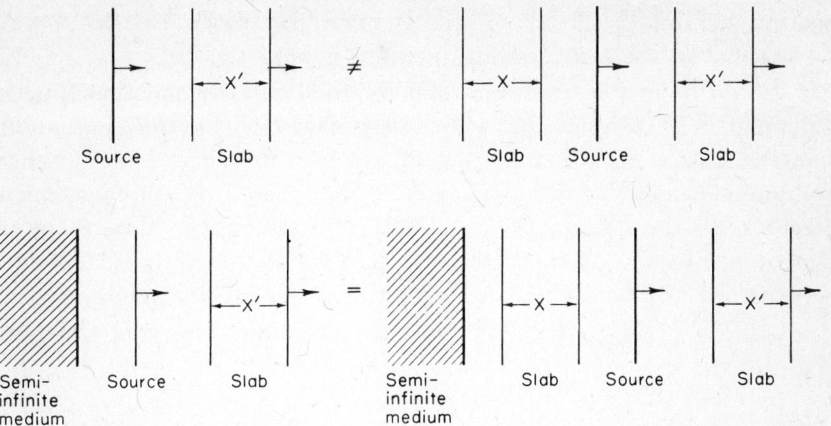

Fig. 2.4 *Functional relation for the modified transmission function.*

2.2.3 Relation between the Modified Transmission Function and the Ordinary Transmission Function

We shall now derive the equation which combines the modified transmission function with the ordinary one.

With the arrangement of a source, slab, and semi-infinite medium given in Fig. 2.3, the incident photons on the slab consist not only of photons emitted directly from the source but also of photons reflected by the semi-infinite medium. Let the intensity of photons emitted directly from the source be $J_0(E, \omega)$ and the incident current on the slab be $J_{\text{in}}(E, \omega)$. Then we have

$$\begin{aligned} \mathbf{J}_{\text{in}} &= \mathbf{J}_0 + \mathbf{R}(\infty)\mathbf{R}(X)\mathbf{J}_0 + \cdots + [\mathbf{R}(\infty)\mathbf{R}(X)]^n\mathbf{J}_0 + \cdots \\ &= [\mathbf{E} - \mathbf{R}(\infty)\mathbf{R}(X)]^{-1}\mathbf{J}_0, \end{aligned} \tag{2.42}$$

where column vectors $\mathbf{J}_0$, etc. are condensed forms of $J_0(E, \omega)$, etc., $\mathbf{E}$ is the unit matrix, and $\mathbf{A}^{-1}$ is the inverse matrix of $\mathbf{A}$.

According to the definition of the transmission functions $\mathbf{T}(X)$ and $\tilde{\mathbf{T}}(X)$, the transmitted current $\mathbf{J}_{\text{tr}}$ is expressed as

$$\begin{aligned} \mathbf{J}_{\text{tr}} &= \mathbf{T}(X)\mathbf{J}_{\text{in}} \\ &= \tilde{\mathbf{T}}(X)\mathbf{J}_0. \end{aligned} \tag{2.43}$$

From Eqs. (2.42) and (2.43), we obtain the formula

$$\mathbf{T}(X) = \tilde{\mathbf{T}}(X)[\mathbf{E} - \mathbf{R}(\infty)\mathbf{R}(X)]. \tag{2.44}$$

When the thickness X is greater than 1 or 2 mean free paths, Eq. (2.43) can be reduced to

$$\mathbf{T}(X) = \tilde{\mathbf{T}}(X)[\mathbf{E} - \mathbf{R}^2(\infty)]. \tag{2.45}$$

The term $\mathbf{R}(\infty)\mathbf{R}(X)$ in Eq. (2.44) represents the effect of the semi-infinite medium behind the source. The magnitude of this term is of the order of the square of the reflection coefficient (albedo) and much smaller than unity except for the case of very oblique incidence.

2.2.4 Flux in an Infinite Homogeneous Medium

Flux in an infinite homogeneous medium can be expressed in terms of the modified transmission function and the reflection function for a semi-infinite medium.

Consider an infinite homogeneous medium with a plane source which emits photons with intensity $J_0^+(E, \omega)$ in the positive x-direction (cf. Fig. 2.5). Let the angular current density in the positive x-direction at

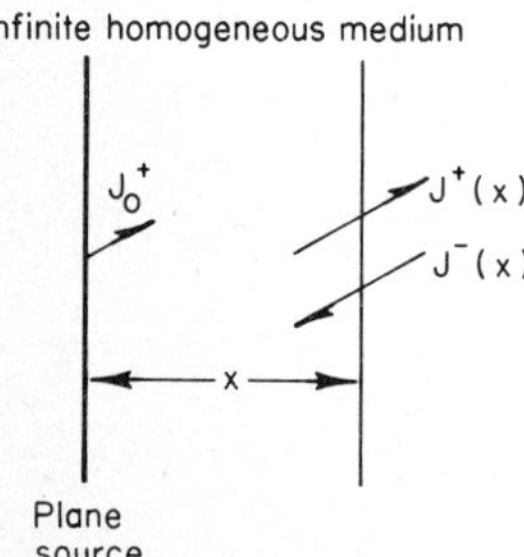

Fig. 2.5 *Currents in an infinite homogeneous medium.*

a distance X from the source be $J^+(X, E, \omega)$ and that in the negative x-direction be $J^-(X, E, \omega)$. Then the following equations hold for these current densities:

$$\mathbf{J}^+(X) = \tilde{\mathbf{T}}(X)\mathbf{J}_0 + \mathbf{R}(\infty)\mathbf{J}^-(X), \tag{2.46}$$

$$\mathbf{J}^-(X) = \mathbf{R}(\infty)\mathbf{J}^+(X). \tag{2.47}$$

By solving Eqs. (2.46) and (2.47), we have

$$\mathbf{J}^{+}(X) + \mathbf{J}^{-}(X) = [\mathbf{E} - \mathbf{R}(\infty)]^{-1}\tilde{\mathbf{T}}(X)\mathbf{J}_0. \tag{2.48}$$

The flux then is given by

$$\phi(X, E) = \int_0^1 \frac{d\omega}{\omega} [J^{+}(X, E, \omega) + J^{-}(X, E, \omega)]. \tag{2.49}$$

When the plane source emits photons with intensity $J_0^{-}(E, \omega)$ in the negative x-direction as well as in the positive x-direction, Eq. (2.48) should be modified to

$$\mathbf{J}^{+}(X) + \mathbf{J}^{-}(X) = [\mathbf{E} - \mathbf{R}(\infty)]^{-1}\tilde{\mathbf{T}}(X)[\mathbf{J}_0^{+} + \mathbf{R}(\infty)\mathbf{J}_0^{-}]. \tag{2.50}$$

The term $[\mathbf{E} - \mathbf{R}(\infty)]^{-1}$ in Eq. (2.48) represents the boundary effect, that is, the difference between the flux in a medium bounded at a distance X from the source and the flux at the same location in a corresponding infinite medium.

In actual computation, it is convenient to introduce the modified current density $\tilde{\mathbf{J}}(X)$ which is defined by the equation

$$\tilde{\mathbf{J}}(X) = \tilde{\mathbf{T}}(X)[\mathbf{J}_0^{+} + \mathbf{R}(\infty)\mathbf{J}_0^{-}]. \tag{2.51}$$

From the functional relation (2.41), it can be shown that the modified current density satisfies the recurrence formula

$$\tilde{\mathbf{J}}(nX + X) = \tilde{\mathbf{T}}(X)\tilde{\mathbf{J}}(nX). \tag{2.52}$$

Equation (2.52) is useful in computing the spatial distribution of flux in infinite homogeneous media or in homogeneous slabs.

CHAPTER THREE *SOLUTIONS OF EQUATIONS FOR SIMPLIFIED MODELS*

Before proceeding to a practical method of solving the equations derived in Chapter 2, we shall solve them for some simplified models. Although solutions obtained for such models are usually too crude to be applicable to actual gamma-ray problems, they may be useful for understanding some qualitative aspects of the basic equations.

A practical method of solving the equations, which is numerical in nature and involves extensive computation, will be given in subsequent chapters.

3.1 One Group Approximation

We shall first adopt a most simple but unrealistic model which assumes that photons moving in a medium have their own energy–angle distribution whose form is kept unchanged at any point in the medium.

With this model, the reflection and transmission functions can be expressed as products of a function of the thickness and a function of the energy and angular variables as

$$T(E, \omega \mid E_0, \omega_0; X) = T(X) f_T(E, \omega \mid E_0, \omega_0), \quad (3.1)$$

$$R(E, \omega \mid E_0, \omega_0; X) = R(X) f_R(E, \omega \mid E_0, \omega_0), \quad (3.2)$$

where f_T and f_R are appropriate functions satisfying the conditions

$$\int_0^\infty dE \int_0^1 d\omega\, f_T(E, \omega \mid E_0, \omega_0) = 1,$$

$$\int_0^\infty dE \int_0^1 d\omega\, f_R(E, \omega \mid E_0, \omega_0) = 1,$$

so that the functions $T(X)$ and $R(X)$ correspond to the total number of transmitted and reflected photons, respectively.

Then, Eqs. (2.19) and (2.20) can be reduced to equations for the scalar functions $T(X)$ and $R(X)$,

$$\frac{d}{dX} R(X) = T(X)\dot{R}T(X), \tag{3.3}$$

$$\frac{d}{dX} T(X) = [R(X)\dot{R} - \dot{T}]T(X). \tag{3.4}$$

The constants $\dot{R}$ and $\dot{T}$ in Eqs. (3.3) and (3.4) correspond to certain averages of the functions $\dot{R}(E, \omega \mid E_0, \omega_0)$ and $\dot{T}(E, \omega \mid E_0, \omega_0)$ given by Eqs. (2.33) and (2.34). We shall, however, not calculate the constants by performing such averages. Instead, we shall treat them as parameters to be adjusted so that the solutions may fit experimental values.

Equations (3.3) and (3.4) can be solved analytically. One obtains

$$T(X) = \frac{K}{K \cosh KX + \dot{T} \sinh KX}, \tag{3.5}$$

$$R(X) = \frac{\dot{R} \sinh KX}{K \cosh KX + \dot{T} \sinh KX}, \tag{3.6}$$

where the constant K is given by

$$K = (\dot{T}^2 - \dot{R}^2)^{1/2}. \tag{3.7}$$

The reflection coefficient for a semi-infinite medium is given by

$$R_\infty = \frac{\dot{R}}{K + \dot{T}}. \tag{3.8}$$

The equation for the modified transmission function becomes

$$\frac{d}{dX} \tilde{T}(X) = -K\tilde{T}(X). \tag{3.9}$$

The solution of Eq. (3.9) is given by

$$\tilde{T}(X) = e^{-KX}. \tag{3.10}$$

The constants $\dot{R}$ and $\dot{T}$ may be determined from measured values of the reflection coefficient R_∞ and the relaxation constant K. By eliminating the constants $\dot{R}$ and $\dot{T}$ using Eqs. (3.7) and (3.8), we have

$$T(X) = \frac{(1 - R_\infty{}^2)e^{-KX}}{1 - R_\infty{}^2 e^{-2KX}}, \tag{3.11}$$

$$R(X) = \frac{R_\infty(1 - e^{-2KX})}{1 - R_\infty{}^2 e^{-2KX}}. \tag{3.12}$$

3.1.1 Dependence of Reflection on Slab Thickness

Fujita *et al.* [30] measured the reflection coefficient (albedo) of iron as a function of slab thickness.

They proposed the experimental formula

$$R(X) = b + (R_\infty - b)(1 - e^{-cX}) \tag{3.13}$$

for slab thickness greater than 0.1 mean free path to fit their experimental results. The values of the constants R_∞, b, and c determined by Fujita *et al.* are given in Table 3.1.

Table 3.1 *Asymptotic Values of Number Albedo and Constants b and c*[a]

Source	Albedo	c (mfp)	b
^{60}Co	0.387 ± 0.01	4.09 ± 0.05	0.037

[a] Obtained experimentally by Fujita *et al.* [30].

Now, by expanding the denominator on the right-hand side of Eq. (3.12), we have

$$R(X) = R_\infty(1 - e^{-2KX}) + R_\infty{}^3 e^{-2KX} - R_\infty{}^3(1 - R_\infty{}^2)e^{-4KX} + \cdots. \tag{3.14}$$

If we neglect third- and higher-order terms on the right-hand side of Eq. (3.14), the resultant expression becomes identical to the experimental formula (3.13) by setting

$$c = 2K, \tag{3.15}$$

$$b = R_\infty{}^3. \tag{3.16}$$

From Table 3.1, we get $R_\infty{}^3 = 0.058$ and $b = 0.037$. Thus, relation (3.16) holds fairly well.

A closer agreement is obtained if we compare the theoretical formula (3.12) directly with the experimental one. The dashed curve in Fig. 3.1 represents the reflection coefficient computed from formula (3.12) using the constants R_∞ and c given in Table 3.1. The solid curve is obtained from the experimental formula (3.13) which contains another parameter b as well as the parameters R_∞ and c.

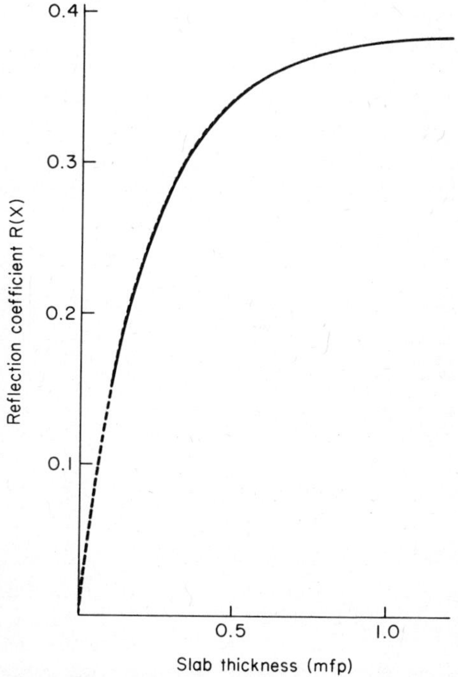

Fig. 3.1 *Number albedo of iron as a function of slab thickness.* — *experimental formula* (3.13), --- *Eq.* (3.12).

3.2 Reflection and Transmission of Monoenergetic Photons by an Isotropically Scattering Medium

When the photon energy is well below 100 keV, its variation due to scattering becomes so small that photons can be treated as monoenergetic. The problems of reflection and transmission of monoenergetic radiations in the case of isotropic, first-order anisotropic, and Rayleigh scattering have been solved extensively by Chandrasekhar [15].

Bellman *et al.* [21] made extensive computations of the reflection function of finite slabs for monoenergetic and isotropically scattered radiations and tabulated the results.

In this section, we shall briefly describe the solutions for monoenergetic and isotropically scattered radiations obtained by Chandrasekhar and Bellman *et al.* More detailed descriptions may be found in the books by Chandrasekhar [15] and Bellman *et al.* [21].

When the scattering is isotropic, the scattering kernel is expressed as

$$\Sigma_s(\omega_0 \to \omega) = \tfrac{1}{2}\tilde{\omega}_0 \Sigma, \tag{3.17}$$

where Σ is the total cross section and $\tilde{\omega}_0$ is the fraction of the scattering per collision given by

$$\tilde{\omega}_0 = \int_{-1}^{1} d\omega \, \Sigma_s(\omega_0 \to \omega)/\Sigma. \tag{3.18}$$

The monoenergetic version of the reflection and transmission functions used in this monograph are somewhat different than the functions introduced by Chandrasekhar. The relation between them is expressed as

$$R(\omega \mid \omega_0; X) = \frac{1}{2\omega_0} S^{(0)}(\tau; \omega, \omega_0), \tag{3.19}$$

$$T(\omega \mid \omega_0; X) = e^{-\tau/\omega}\, \delta(\omega - \omega_0) + \frac{1}{2\omega_0} T^{(0)}(\tau; \omega, \omega_0), \tag{3.20}$$

where $S^{(0)}(\tau; \omega, \omega_0)$ and $T^{(0)}(\tau; \omega, \omega_0)$ are Chandrasekhar's scattering and transmission functions, respectively,* and $\tau = \Sigma X$.

* In the fields of astrophysics and neutron physics, the cosine of obliquity is customarily denoted by μ. We, however, use the symbol ω instead of μ, since μ is usually used to represent the mass absorption coefficient in the field of gamma-ray physics.

They are expressed in terms of more general scattering and transmission functions involving the azimuthal angles introduced by Chandrasekhar as

$$S^{(0)}(\tau;\omega,\omega_0)=\frac{1}{2\pi}\int_0^{2\pi}d(\varphi-\varphi_0)S(\tau;\omega,\varphi;\omega_0,\varphi_0),$$

$$T^{(0)}(\tau;\omega,\omega_0)=\frac{1}{2\pi}\int_0^{2\pi}d(\varphi-\varphi_0)T(\tau;\omega,\varphi;\omega_0,\varphi_0).$$

The functions satisfy the reciprocity law

$$S^{(0)}(\tau;\omega,\omega_0)=S^{(0)}(\tau;\omega_0,\omega), \tag{3.21}$$

$$T^{(0)}(\tau;\omega,\omega_0)=T^{(0)}(\tau;\omega_0,\omega). \tag{3.22}$$

The proof of the principle of reciprocity was given by Chandrasekhar based on the fact that the scattering kernel satisfies the condition

$$\Sigma_s(\omega_0\to\omega)=\Sigma_s(\omega\to\pm\omega_0). \tag{3.23}$$

The reciprocity is due to the fact that the motion of a monoenergetic photon between any two points in a medium is reversible. This is true only for monoenergetic photons. For the case in which the photon energy is reduced by scattering, the reciprocity law does not hold unless the cross section is independent of the energy.

3.2.1 Reflection Function for Semi-Infinite Media

In the case of isotropic scattering, the scattering function for a semi-infinite medium satisfies the equation

$$\begin{aligned}\left(\frac{1}{\omega}+\frac{1}{\omega_0}\right)&S(\omega,\omega_0)\\ &=\tilde{\omega}_0\left[1+\frac{1}{2}\int_0^1 S(\omega,\omega')\frac{d\omega'}{\omega'}+\frac{1}{2}\int_0^1 S(\omega'',\omega_0)\frac{d\omega''}{\omega''}\right.\\ &\left.\quad+\frac{1}{4}\int_0^1\int_0^1 S(\omega,\omega')S(\omega'',\omega_0)\frac{d\omega'}{\omega'}\frac{d\omega''}{\omega''}\right],\end{aligned} \tag{3.24}$$

where $S(\omega,\omega_0)$ is an abbreviated expression for $S^{(0)}(\infty;\omega,\omega_0)$.

Since the function $S(\omega, \omega_0)$ is symmetrical with respect to ω and ω_0, Eq. (3.24) can be written as

$$\left(\frac{1}{\omega} + \frac{1}{\omega_0}\right) S(\omega, \omega_0) = \tilde{\omega}_0 H(\omega) H(\omega_0), \tag{3.25}$$

where

$$H(\omega) = 1 + \frac{1}{2} \int_0^1 S(\omega, \omega') \frac{d\omega'}{\omega'}. \tag{3.26}$$

From Eqs. (3.25) and (3.26), one obtains the equation for the function $H(\omega)$,

$$H(\omega) = 1 + \frac{1}{2} \tilde{\omega}_0 \omega H(\omega) \int_0^1 \frac{H(\omega')}{\omega + \omega'} d\omega'. \tag{3.27}$$

Thus, the scattering function is expressed in terms of the function $H(\omega)$, called the H function.

The properties of the H function as well as its numerical values were given by Chandrasekhar [15].

It was shown that an integral of the H function for isotropic scattering is expressed as

$$\int_0^1 H(\omega)\, d\omega = \frac{2}{\tilde{\omega}_0} [1 - (1 - \tilde{\omega}_0)^{1/2}]. \tag{3.28}$$

From Eqs. (3.19) and (3.25), one finally obtains

$$R(\omega \mid \omega_0; \infty) = \frac{\tilde{\omega}_0 \omega H(\omega) H(\omega_0)}{2(\omega + \omega_0)}.$$

When the incident current is isotropic, we have

$$\begin{aligned} R(\omega) &= \int_0^1 d\omega_0\, R(\omega \mid \omega_0; \infty) \\ &= H(\omega) - 1, \end{aligned} \tag{3.29}$$

where we have used Eq. (3.27) to derive Eq. (3.29).

In this case the total number of reflected photons is expressed as

$$\begin{aligned} R &= \int_0^1 d\omega\, R(\omega) \\ &= \frac{2}{\tilde{\omega}_0} [1 - (1 - \tilde{\omega}_0)^{1/2}] - 1. \end{aligned} \tag{3.30}$$

It is interesting to compare the number of reflected photons given by Eq. (3.30) with that of singly scattered photons, which is readily found to be given by $\tilde{\omega}_0/4$.

The number of reflected photons in the case of isotropic incidence is plotted in Fig. 3.2. It is seen that the fraction of singly scattered photons decreases to $\frac{1}{4}$ as the parameter $\tilde{\omega}_0$ approaches unity.

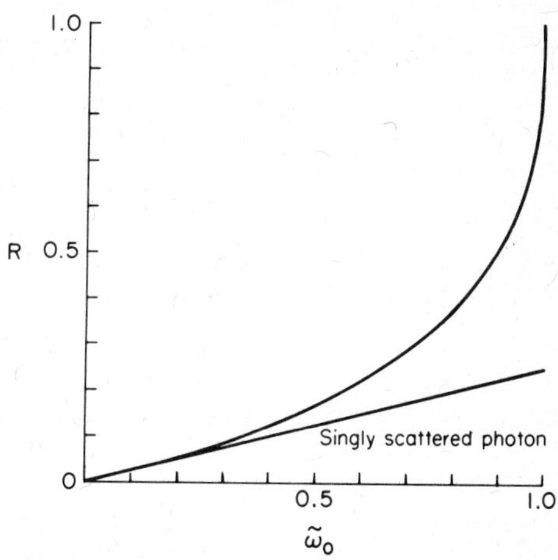

Fig. 3.2 *Albedo for monoenergetic and isotropically scattered radiation. Abscissa is the fraction of scattering per collision.*

3.2.2 Direct Numerical Solution of $S(\omega, \omega_0)$

The reflection function for a semi-infinite medium $S(\omega, \omega_0)$ can also be obtained by direct numerical solution of Eq. (3.24).

In solving the equation numerically, we shall replace the integration over the angular variable in Eq. (3.24) by the summation over discrete numbers by using the Gaussian quadrature formula in the range $0 \leq \omega \leq 1$:

$$\int_0^1 g(\omega)\, d\omega = \sum_{k=1}^{G} C_k g(\omega_k), \tag{3.31}$$

where ω_k is the kth point of division, C_k the corresponding weight, and G the total number of divisions.

Then, Eq. (3.24) can be reduced approximately to

$$\left(\frac{1}{\omega_i}+\frac{1}{\omega_j}\right)S(\omega_i,\omega_j)=\tilde{\omega}_0\left\{1+\frac{1}{2}\sum_{k=1}^{G}\frac{C_k}{\omega_k}S(\omega_i,\omega_k)\right\}\times\left\{1+\frac{1}{2}\sum_{k=1}^{G}\frac{C_k}{\omega_k}S(\omega_k,\omega_j)\right\}. \tag{3.32}$$

Equation (3.32) was solved by an iteration method.

The accuracy of numerical solutions as a function of the total number of divisions G was examined.

In Table 3.2, the total number of reflected photons for an isotropic incident current evaluated by numerical solution is compared with the exact value given by Eq. (3.30). The numerical solutions obtained with $G = 3$ are accurate enough for practical applications in the case of isotropic scattering.

Table 3.2 *Albedo of Monoenergetic and Isotropically Scattered Radiations*[a]

$\tilde{\omega}_0$	Exact	Gaussian approximations ($G = 3$)	Number of iterations
0.1	0.026334	0.026334	3
0.3	0.0889331	0.088933	5
0.5	0.1715729	0.17157	7
0.7	0.292213	0.29221	9
0.9	0.519494	0.51943	17
0.95	0.634512	0.63439	24
1.00	1.000000	0.9425	50[b]

[a] The incident current is isotropic.

[b] The computation is given up at 50 iterations.

3.2.3 Reflection and Transmission Functions for Slabs of Finite Thickness

Since the scattering function for semi-infinite medium is expressed with the H function, the scattering and transmission functions for slabs of finite thickness can also be expressed in terms of the X and Y functions in the forms

$$\left(\frac{1}{\omega_0}+\frac{1}{\omega}\right)S^{(0)}(\tau;\,\omega,\,\omega_0)=\tilde{\omega}_0[X(\omega,\,\tau)X(\omega_0,\,\tau)-Y(\omega,\,\tau)Y(\omega_0,\,\tau)], \tag{3.33}$$

$$\left(\frac{1}{\omega_0}-\frac{1}{\omega}\right)T^{(0)}(\tau;\,\omega,\,\omega_0)=\tilde{\omega}_0[Y(\omega,\,\tau)X(\omega_0,\,\tau)-X(\omega,\,\tau)Y(\omega_0,\,\tau)]. \tag{3.34}$$

The X and Y functions are defined as the solutions of the integral equations

$$X(\omega,\,\tau)=1+\tfrac{1}{2}\tilde{\omega}_0\omega\int_0^1\frac{d\omega'}{\omega+\omega'}\,[X(\omega,\,\tau)X(\omega',\,\tau)-Y(\omega,\,\tau)Y(\omega',\,\tau)], \tag{3.35}$$

$$Y(\omega,\,\tau)=e^{-\tau/\omega}+\tfrac{1}{2}\tilde{\omega}_0\omega\int_0^1\frac{d\omega'}{\omega-\omega'}\,[Y(\omega,\,\tau)X(\omega',\,\tau) - X(\omega,\,\tau)Y(\omega',\,\tau)], \tag{3.36}$$

satisfying the conditions

$$X(\omega,\,\tau)\to H(\omega) \quad \text{and} \quad Y(\omega,\,\tau)\to 0 \quad \text{as} \quad \tau\to\infty \tag{3.37}$$

and

$$X(\omega,\,\tau)\to 1 \quad \text{and} \quad Y(\omega,\,\tau)\to e^{-\tau/\omega} \quad \text{as} \quad \tau\to 0. \tag{3.38}$$

Several properties of the X and Y functions were given by Chandrasekhar [15].

3.2.4 Direct Numerical Integration of the Equation for the Reflection Function

Bellman *et al.* [21] have made extensive calculations of the reflection functions for monoenergetic and isotropically scattered radiations.

The solutions were obtained by direct numerical integration of the equation with the Gaussian quadrature formula (3.31).

By means of Eq. (3.31), the equation for the scattering function is reduced to

$$\frac{\partial}{\partial \tau} S^{(0)}(\tau; \omega_i, \omega_j) = -\left(\frac{1}{\omega_i} + \frac{1}{\omega_j}\right) S^{(0)}(\tau; \omega_i, \omega_j)$$
$$+ \tilde{\omega}_0 \left\{1 + \frac{1}{2} \sum_{k=1}^{G} \frac{C_k}{\omega_k} S^{(0)}(\tau; \omega_i, \omega_k)\right\}$$
$$\times \left\{1 + \frac{1}{2} \sum_{k=1}^{G} \frac{C_k}{\omega_k} S^{(0)}(\tau, \omega_k, \omega_j)\right\}. \qquad (3.39)$$

Equation (3.39) was integrated over a spatial variable τ by the Runge–Kutta method.

Actual calculations were performed using $G = 7$ [G is the total number of divisions in formula (3.31)]. Numerical solutions have been tabulated by Bellman *et al.* [21].

CHAPTER FOUR *METHOD OF NUMERICAL SOLUTIONS*

In this chapter we shall solve the equations for the reflection and transmission functions based on the realistic cross section for gamma rays.

The extension of the method of solution from the problem of light to that of gamma rays is not straightforward. The problem of light is a one-velocity problem with nearly isotropic scattering. On the other hand, the problem of gamma rays is an energy-dependent problem. The variation of energy in Compton scattering has a major effect on the penetration of gamma rays. The scattering of gamma rays is highly anisotropic and is complicated by the energy–angle correlation.

The complicated forms of the scattering kernel and the total cross section rule out analytical solutions as obtained by Chandrasekhar [15] for the problem of light. Solutions for gamma rays have been obtained by the numerical solution of the equations with multigroup approximation for energy and the Gaussian approximation for the angular variable.

4.1 Derivation of Equations for Numerical Solutions

The transmission function and the modified transmission function can respectively be expressed by a sum of two terms as

$$T(E, \omega \mid E_0, \omega_0; X) = e^{-\Sigma(E)X/\omega}\,\delta(E - E_0)\,\delta(\omega - \omega_0) + T^{(s)}(E, \omega \mid E_0, \omega_0; X), \tag{4.1}$$

$$\tilde{T}(E, \omega \mid E_0, \omega_0; X) = e^{-\Sigma(E)X/\omega}\,\delta(E - E_0)\,\delta(\omega - \omega_0) + \tilde{T}^{(s)}(E, \omega \mid E_0, \omega_0; X). \tag{4.2}$$

The first terms in Eqs. (4.1) and (4.2) represent unscattered photons and the second terms scattered ones.

In Compton scattering, photons are invariably scattered into lower energy. Therefore, the functions $T^{(s)}(E, \omega \mid E_0, \omega_0; X)$, $\tilde{T}^{(s)}(E, \omega \mid E_0, \omega_0; X)$, and $R(E, \omega \mid E_0, \omega_0; X)$ vanish for energy E greater than source energy E_0. From Eqs. (4.2) and (2.37), we obtain the equation for the modified transmission function for scattered photons,

$$\begin{aligned} \frac{\partial}{\partial X}\tilde{T}^{(s)}(E, \omega \mid E_0, \omega_0; X) &= -\frac{\Sigma(E)}{\omega}\tilde{T}^{(s)}(E, \omega \mid E_0, \omega_0; X) \\ &\quad + \int_0^{E_0} dE' \int_0^1 d\omega'\, C(E, \omega \mid E', \omega')\tilde{T}^{(s)}(E', \omega' \mid E_0, \omega_0; X) \\ &\quad + C(E, \omega \mid E_0, \omega_0)\exp\Big(-\frac{\Sigma(E_0)}{\omega_0}X\Big). \end{aligned} \tag{4.3}$$

The initial condition for the function is evidently given by

$$\tilde{T}^{(s)}(E, \omega \mid E_0, \omega_0; 0) = 0. \tag{4.4}$$

In terms of the function $\tilde{T}^{(s)}$, the functional relation (2.40) can be rewritten in the form

$$\begin{aligned}
&\tilde{T}^{(s)}(E, \omega \mid E_0, \omega_0; X + X') \\
&\quad = \int_0^{E_0} dE' \int_0^1 d\omega' \, \tilde{T}^{(s)}(E, \omega \mid E', \omega'; X')\tilde{T}^{(s)}(E', \omega' \mid E_0, \omega_0; X) \\
&\quad + \exp\left(-\frac{\Sigma(E)}{\omega} X'\right) \cdot \tilde{T}^{(s)}(E, \omega \mid E_0, \omega_0; X) \\
&\quad + \tilde{T}^{(s)}(E, \omega \mid E_0, \omega_0; X') \exp\left(-\frac{\Sigma(E_0)}{\omega_0} X\right).
\end{aligned} \tag{4.5}$$

The relation between the modified transmission function and the ordinary one [Eq. (2.44)] is expressed as

$$\begin{aligned}
&T^{(s)}(E, \omega \mid E_0, \omega_0; X) \\
&\quad = \tilde{T}^{(s)}(E, \omega \mid E_0, \omega_0; X) - \exp\left(-\frac{\Sigma(E)}{\omega} X\right) A(E, \omega \mid E_0, \omega_0; X) \\
&\quad - \int_0^{E_0} dE' \int_0^1 d\omega' \, \tilde{T}^{(s)}(E, \omega \mid E', \omega'; X) A(E', \omega' \mid E_0, \omega_0; X),
\end{aligned} \tag{4.6}$$

where

$$\begin{aligned}
&A(E, \omega \mid E_0, \omega_0; X) \\
&\quad = \int_0^{E_0} dE' \int_0^1 d\omega' \, R(E, \omega \mid E', \omega'; \infty) R(E', \omega' \mid E_0, \omega_0; X).
\end{aligned} \tag{4.7}$$

4.1.1 Multigroup Approximation

We divide the energy range of interest (usually from about 0.01 to 10 MeV) into a number of groups, that is N groups, and introduce the functions from the group m to n defined by the equations

$$\begin{aligned}
R_{nm}(\omega, \omega_0; X) &= \frac{1}{\Delta E_m} \int_n dE \int_m dE_0 \, R(E, \omega \mid E_0, \omega_0; X), \\
T^{(s)}_{nm}(\omega, \omega_0; X) &= \frac{1}{\Delta E_m} \int_n dE \int_m dE_0 \, T^{(s)}(E, \omega \mid E_0, \omega_0; X), \qquad (4.8) \\
\tilde{T}^{(s)}_{nm}(\omega, \omega_0; X) &= \frac{1}{\Delta E_m} \int_n dE \int_m dE_0 \, \tilde{T}^{(s)}(E, \omega \mid E_0, \omega_0; X).
\end{aligned}$$

Here, $\int_n dE$ represents the integration over the group n and ΔE_m the width of the group m. Assuming that the energy groups are numbered in the sequence from the highest to the lowest energy, the function R_{nm}, $T^{(s)}_{nm}$, and $\tilde{T}^{(s)}_{nm}$ vanish unless n is equal to or greater than m.

We shall now adopt the following approximation:

A. The integration over energy in the equations is approximated by a summation over the groups with the assumption:

For E within the group n and E_0 within the group m,

$$\begin{aligned} R(E, \omega \mid E_0, \omega_0; X) &\simeq \text{constant} = \frac{1}{\Delta E_n} R_{nm}(\omega, \omega_0; X), \\ T^{(s)}(E, \omega \mid E_0, \omega_0; X) &\simeq \text{constant} = \frac{1}{\Delta E_n} T^{(s)}_{nm}(\omega, \omega_0; X), \\ \tilde{T}^{(s)}(E, \omega \mid E_0, \omega_0; X) &\simeq \text{constant} = \frac{1}{\Delta E_n} \tilde{T}^{(s)}_{nm}(\omega, \omega_0; X). \end{aligned} \tag{4.9}$$

With this approximation, the equations for the reflection function of a semi-infinite medium and the modified transmission function [Eqs. (2.35) and (4.3)] can be reduced to

$$\begin{aligned} \left[\frac{\Sigma_n}{\omega} + \frac{\Sigma_m}{\omega_0}\right] & R_{nm}(\omega, \omega_0; \infty) \\ = {} & \frac{1}{\omega_0} \Sigma_{nm}(-\omega, \omega_0) \\ & + \sum_{l=m}^{n} \int_0^1 \frac{d\omega'}{\omega'} \Sigma_{nl}(\omega, \omega') R_{lm}(\omega', \omega_0; \infty) \\ & + \frac{1}{\omega_0} \sum_{l=m}^{n} \int_0^1 d\omega' \, R_{nl}(\omega, \omega'; \infty) \Sigma_{lm}(\omega', \omega) \\ & + \sum_{l=m}^{n} \int_0^1 d\omega' \, R_{nl}(\omega, \omega'; \infty) \\ & \times \sum_{s=m}^{l} \int_0^1 \frac{d\omega''}{\omega''} \Sigma_{ls}(-\omega', \omega'') R_{sm}(\omega'', \omega_0; \infty), \end{aligned} \tag{4.10}$$

and

$$\begin{aligned} \frac{\partial}{\partial X} & \tilde{T}^{(s)}_{nm}(\omega, \omega_0; X) \\ & = -\frac{\Sigma_n}{\omega} \tilde{T}^{(s)}_{nm}(\omega, \omega_0; X) + \sum_{l=m}^{n} \int_0^1 d\omega' \, C_{nl}(\omega, \omega') \tilde{T}^{(s)}_{lm}(\omega', \omega; X) \\ & \quad + C_{nm}(\omega, \omega_0) \exp\left(-\frac{\Sigma_m}{\omega_0} X\right). \end{aligned} \tag{4.11}$$

Here, Σ_n is the average total cross section of the group n defined by

$$\Sigma_n = \frac{1}{\Delta E_n} \int_n dE\, \Sigma(E), \tag{4.12}$$

and $\Sigma_{nm}(\omega, \omega_0)$ is the differential cross section from the group m to n and from the direction ω_0 to ω given by

$$\Sigma_{nm}(\omega, \omega_0) = \frac{1}{\Delta E_m} \int_n dE \int_m dE_0\, \Sigma_s(E_0, \omega_0 \to E, \omega). \tag{4.13}$$

The function $C_{nm}(\omega, \omega_0)$ in Eq. (4.11) is defined by

$$C_{nm}(\omega, \omega_0) = \frac{1}{\Delta E_m} \int_n dE \int_m dE_0\, C(E, \omega \mid E_0, \omega_0), \tag{4.14}$$

where $C(E, \omega \mid E_0, \omega_0)$ is given by Eq. (2.39). For the multigroup approximation, we have

$$C_{nm}(\omega, \omega_0) = \frac{1}{\omega_0} \Sigma_{nm}(\omega, \omega_0) + \frac{1}{\omega_0} \sum_{l=m}^{n} \int_0^1 d\omega'\, R_{nl}(\omega, \omega'; \infty) \Sigma_{lm}(-\omega', \omega_0). \tag{4.15}$$

4.1.2 Gaussian Approximation

We further adopt the following approximation:

B. The integration over the angular variable in the equations is approximated by finite sums by means of the Gaussian quadrature in the range $0 \leq \omega \leq 1$,

$$\int_0^1 g(\omega)\, d\omega \simeq \sum_{k=1}^{G} C_k g(\omega_k), \tag{4.16}$$

where ω_k is the kth point of division, C_k the corresponding weight, and G the total number of divisions. For the Gaussian approximation, Eqs.

(4.10) and (4.11) will be reduced to

$$\left(\frac{\Sigma_n}{\omega_i}+\frac{\Sigma_m}{\omega_j}\right)R_{nm}(\omega_i,\omega_j;\infty)$$
$$=\frac{1}{\omega_j}\Sigma_{nm}(-\omega_i,\omega_j)$$
$$+\sum_{l=m}^{n}\sum_{k=1}^{G}\frac{C_k}{\omega_k}\Sigma_{nl}(\omega_i,\omega_k)R_{lm}(\omega_k,\omega_j;\infty)$$
$$+\frac{1}{\omega_j}\sum_{l=m}^{n}\sum_{k=1}^{G}C_kR_{nl}(\omega_i,\omega_k;\infty)\Sigma_{lm}(\omega_k,\omega_j)$$
$$+\sum_{l=m}^{n}\sum_{k=1}^{G}C_kR_{nl}(\omega_i,\omega_k;\infty)$$
$$\times\left[\sum_{s=m}^{l}\sum_{p=1}^{G}\frac{C_p}{\omega_p}\Sigma_{ls}(-\omega_k,\omega_p)R_{sm}(\omega_p,\omega_j;\infty)\right],\qquad 1\le i,\,j\le G, \tag{4.17}$$

$$\frac{\partial}{\partial X}\tilde{T}^{(s)}_{nm}(\omega_i,\omega_j;X)$$
$$=-\frac{\Sigma_n}{\omega_i}\tilde{T}^{(s)}_{nm}(\omega_i,\omega_j;X)$$
$$+\sum_{l=m}^{n}\sum_{k=1}^{G}C_kC_{nl}(\omega_i,\omega_k)\tilde{T}^{(s)}_{lm}(\omega_k,\omega_j;X)$$
$$+C_{nm}(\omega_i,\omega_j)\exp\left(-\frac{\Sigma_m}{\omega_j}X\right),\qquad 1\le i,\,j\le G. \tag{4.18}$$

Here, the function $C_{nm}(\omega_i,\omega_j)$ is given by the equation

$$C_{nm}(\omega_i,\omega_j)=\frac{1}{\omega_j}\Sigma_{nm}(\omega_i,\omega_j)$$
$$+\frac{1}{\omega_j}\sum_{l=m}^{n}\sum_{k=1}^{G}C_kR_{nl}(\omega_i,\omega_k;\infty)\Sigma_{lm}(-\omega_k,\omega_j). \tag{4.19}$$

Equations (4.17) and (4.18) are our basic equations for numerical solutions. The initial condition for the function $\tilde{T}^{(s)}_{nm}(\omega_i,\omega_j;X)$ is evidently given by

$$\tilde{T}^{(s)}_{nm}(\omega_i,\omega_j;0)=0. \tag{4.20}$$

The functional relation for the function $\tilde{T}^{(s)}_{nm}(\omega_i,\omega_j;X)$ corresponding to Eq. (4.5) can be written in the form

$$\tilde{T}^{(s)}_{nm}(\omega_i, \omega_j; X + X') = \sum_{l=m}^{n} \sum_{k=1}^{G} C_k \tilde{T}^{(s)}_{nl}(\omega_i, \omega_k; X') \tilde{T}^{(s)}_{lm}(\omega_k, \omega_j; X)$$
$$+ \exp\left(-\frac{\Sigma_n}{\omega_i} X'\right) \tilde{T}^{(s)}_{nm}(\omega_i, \omega_j; X)$$
$$+ \tilde{T}^{(s)}_{nm}(\omega_i, \omega_j; X') \exp\left(-\frac{\Sigma_m}{\omega_j} X\right). \qquad (4.21)$$

It can be proved that the solution of Eq. (4.19) with the initial condition (4.20) satisfies rigorously the functional relation (4.21).

Corresponding to Eq. (4.6), which expresses the relation between the modified transmission function and the ordinary one, we have

$$T^{(s)}_{nm}(\omega_i, \omega_j; X) = \tilde{T}^{(s)}_{nm}(\omega_i, \omega_j; X) - \exp\left(-\frac{\Sigma_n}{\omega_i} X\right) A_{nm}(\omega_i, \omega_j; X)$$
$$- \sum_{l=m}^{n} \sum_{k=1}^{G} C_k \tilde{T}^{(s)}_{nl}(\omega_i, \omega_k; X) A_{lm}(\omega_k, \omega_j; X), \qquad (4.22)$$

where

$$A_{nm}(\omega_i, \omega_j) = \sum_{l=m}^{n} \sum_{k=1}^{G} C_k R_{nl}(\omega_i, \omega_k; \infty) R_{lm}(\omega_k, \omega_j; X). \qquad (4.23)$$

4.1.3 Energy Reflection and Transmission Functions

In shielding designs it is often necessary to compute the ratio of the energy of reflected or transmitted photons to the energy of incident photons. For this purpose, it is more convenient to introduce the energy reflection and transmission functions for multigroup approximations.

They are defined by the equations

$$R_{\mathrm{E}nm}(\omega, \omega_0; X) = \frac{1}{\Delta E_m} \int_n dE \int_m dE_0 \frac{E}{E_0} R(E, \omega \mid E_0, \omega_0; X),$$
$$T^{(s)}_{\mathrm{E}nm}(\omega, \omega_0; X) = \frac{1}{\Delta E_m} \int_n dE \int_m dE_0 \frac{E}{E_0} T^{(s)}(E, \omega \mid E_0, \omega_0; X), \qquad (4.24)$$
$$\tilde{T}^{(s)}_{\mathrm{E}nm}(\omega, \omega_0; X) = \frac{1}{\Delta E_m} \int_n dE \int_m dE_0 \frac{E}{E_0} \tilde{T}^{(s)}(E, \omega \mid E_0, \omega_0; X).$$

The equations for the energy reflection and transmission functions have the same form as Eqs. (4.17)–(4.23), provided the scattering kernel $\Sigma_{nm}(\omega, \omega_0)$ in these equations is replaced by the energy transfer kernel given by

$$\Sigma_{\mathrm{E}nm}(\omega, \omega_0) = \frac{1}{\Delta E_m} \int_n dE \int_m dE_0 \frac{E}{E_0} \Sigma_s(E_0, \omega_0 \to E, \omega). \quad (4.25)$$

4.2 Method for Solution of Numerical Equations

4.2.1 Reflection Function for an Infinite Medium

Because of the property of the reflection function that its (n, m) component R_{nm} vanishes for $n < m$, the right-hand side of Eq. (4.17) involves only the components R_{nl} and R_{ln}, $m \leq l \leq n$. Therefore, the equation for a diagonal component R_{nn} does not contain other components. It is written in the form

$$\begin{aligned}
\left(\frac{\Sigma_n}{\omega_i} + \frac{\Sigma_n}{\omega_j}\right) & R_{nn}(\omega_i, \omega_j; \infty) \\
&= \frac{1}{\omega_j} \Sigma_{nn}(-\omega_i, \omega_j) \\
&\quad + \sum_{k=1}^{G} \frac{C_k}{\omega_k} \Sigma_{nn}(\omega_i, \omega_k) R_{nn}(\omega_k, \omega_j; \infty) \\
&\quad + \frac{1}{\omega_j} \sum_{k=1}^{G} C_k R_{nn}(\omega_i, \omega_k; \infty) \Sigma_{nn}(\omega_k, \omega_j) \\
&\quad + \sum_{k=1}^{G} C_k R_{nn}(\omega_i, \omega_k; \infty) \\
&\quad \times \sum_{p=1}^{G} \frac{C_p}{\omega_p} \Sigma_{nn}(-\omega_k, \omega_p) R_{nn}(\omega_p, \omega_j; \infty), \qquad 1 \leq i, j \leq G.
\end{aligned} \quad (4.26)$$

The equation for a nondiagonal element R_{nm} $(n > m)$ can be rewritten as

$$\left(\frac{\Sigma_n}{\omega_i} + \frac{\Sigma_m}{\omega_j}\right) R_{nm}(\omega_i, \omega_j) \\ = Q_{nm}(\omega_i, \omega_j) \\ + \sum_{k=1}^{G} C_k [A_n(\omega_i, \omega_k) R_{nm}(\omega_k, \omega_j) + R_{nm}(\omega_i, \omega_k) B_m(\omega_k, \omega_j)]. \tag{4.27}$$

Here, $R_{nm}(\omega_i, \omega_j)$ is an abbreviation of $R_{nm}(\omega_i, \omega_j; \infty)$ and the functions $A_n(\omega_i, \omega_j)$, $B_n(\omega_i, \omega_j)$, and $Q_{nm}(\omega_i, \omega_j)$ are given by the equations

$$A_n(\omega_i, \omega_j) = \frac{1}{\omega_j} \left[\Sigma_{nn}(\omega_i, \omega_j) + \sum_{p=1}^{G} R_{nn}(\omega_i, \omega_p) C_p \Sigma_{nn}(-\omega_p, \omega_j) \right], \tag{4.28}$$

$$B_n(\omega_i, \omega_j) = \frac{1}{\omega_j} \Sigma_{nn}(\omega_i, \omega_j) + \sum_{p=1}^{G} \frac{C_p}{\omega_p} \Sigma_{nn}(-\omega_i, \omega_p) R_{nn}(\omega_p, \omega_j), \tag{4.29}$$

$$Q_{nm}(\omega_i, \omega_j) = \frac{1}{\omega_j} \Sigma_{nm}(\omega_i, \omega_j) + \Bigg[\sum_{l=m}^{n} \sum_{k=1}^{G} \frac{C_k}{\omega_k} \Sigma_{nl}(\omega_i, \omega_k) R_{lm}(\omega_k, \omega_j) \\ + \frac{1}{\omega_j} \sum_{l=m}^{n} \sum_{k=1}^{G} C_k R_{nl}(\omega_i, \omega_k) \Sigma_{lm}(\omega_k, \omega_j) \\ + \sum_{l=m}^{n} \sum_{s=m}^{l} \sum_{k=1}^{G} C_k R_{nl}(\omega_i, \omega_k) \\ \times \sum_{p=1}^{G} \frac{C_p}{\omega_p} \Sigma_{ls}(-\omega_k, \omega_p) R_{sm}(\omega_p, \omega_j) \Bigg]_{R_{nm}=0}. \tag{4.30}$$

If we solve the equations for these components in the sequence R_{11}, R_{22}, R_{21}, ..., R_{nn}, $R_{n\,n-1}$, ..., $R_{n\,m+1}$, R_{nm}, ... as shown in Fig. 4.1, the functions A_n, B_m, and Q_{nm} in Eq. (4.26) may be considered as given on solving the equation for R_{nm}.

Thus, Eq. (4.17) has been reduced to a sequence of integral equations which involves G^2 unknown quantities, where G is the total number of angular divisions.

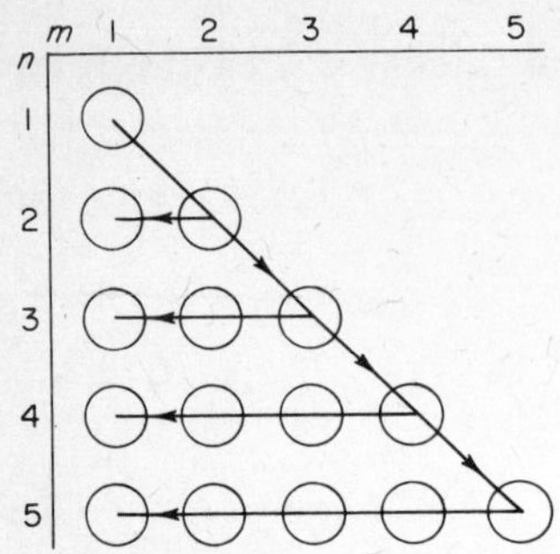

Fig. 4.1 *Sequence in solving R_{nm}* [22].

The integral equations (4.26) and (4.27) were solved by an iteration method [22]. The first-order solutions in the iteration are chosen as

$$R_{nn}^{(1)}(\omega_i, \omega_j; \infty) = \frac{1}{\omega_j} \Sigma_{nn}(-\omega_i, \omega_j) \Big/ \left(\frac{\Sigma_n}{\omega_i} + \frac{\Sigma_n}{\omega_j}\right),$$

$$R_{nm}^{(1)}(\omega_i, \omega_j; \infty) = Q_{nm}(\omega_i, \omega_j) \Big/ \left(\frac{\Sigma_n}{\omega_i} + \frac{\Sigma_m}{\omega_j}\right).$$

The first-order solution for the diagonal component corresponds to the reflection function for singly scattered photons. The first-order solution for the nondiagonal component involves some multiply scattered photons as well as singly scattered photons. Since the contribution of multiply scattered gamma rays to the reflection function is not very large, solutions converge rapidly with iterations. In most cases, solutions converge in 2 to 5 iterations.

4.2.2 Modified Transmission Function

On solving Eq. (4.18), it should be noted that if one obtains the solution for an arbitrary thickness h, one can compute the functions for thicknesses $2h$, $4h$, . . . , $2^n h$, and $3h$, $4h$, . . . , nh, . . . by successive applications of the functional relation (4.21). Therefore, we have only to solve Eq. (4.18) for an initial thickness h which can be chosen as small as desired. It should, however, be noted that the solution for the given initial thickness must be accurate enough to be used to generate solutions for other thicknesses by applying the functional relation.

We shall first solve Eq. (4.18) for an initial thickness h by the Runge–Kutta method.

Equation (4.18) can be rewritten as

$$\frac{\partial}{\partial X} \tilde{T}^{(s)}_{nm}(\omega_i, \omega_j; X) = -\alpha_i \tilde{T}^{(s)}_{nm}(\omega_i, \omega_j; X) + \sum_{k \neq i} C_k C_{nm}(\omega_i, \omega_k) \tilde{T}^{(s)}_{nm}(\omega_k, \omega_j; X) + P_{nm}(\omega_i, \omega_j; X), \qquad 1 \leq i, j \leq G, \tag{4.31}$$

where

$$\alpha_i = \frac{\Sigma_n}{\omega_i} - C_i C_{nm}(\omega_i, \omega_i), \tag{4.32}$$

$$P_{nm}(\omega_i, \omega_j; X) = \sum_{l=m}^{n-1} \sum_{k=1}^{G} C_k C_{nl}(\omega_i, \omega_k) \tilde{T}^{(s)}_{lm}(\omega_k, \omega_j; X) + C_{nm}(\omega_i, \omega_j) \exp\left(-\frac{\Sigma_m}{\omega_j} X\right). \tag{4.33}$$

If we solve the equations for the components of the function in the sequence shown in Fig. 4.1, the function P_{nm} may be considered as given on solving the equation for $\tilde{T}^{(s)}_{nm}$ [Eq. (4.31)].

If we write

$$\tilde{T}^{(s)}_{nm}(\omega_i, \omega_j; X) = \exp(-\alpha_i X) Z_{nm}(\omega_i, \omega_j; X), \tag{4.34}$$

then Eq. (4.31) is reduced to

$$\frac{\partial}{\partial X} Z_{nm}(\omega_i, \omega_j; X) = \sum_{k \neq i} C_k C_{nm}(\omega_i, \omega_k) \exp[(\alpha_i - \alpha_k) X] Z_{nm}(\omega_k, \omega_j; X) + \exp(\alpha_i X) P_{nm}(\omega_i, \omega_j; X). \tag{4.35}$$

Equation (4.35) was solved by the Runge–Kutta method for an initial thickness h. The error involved in the solution is less than h^4, if the initial thickness is chosen sufficiently small.

From Eq. (4.21) we have

$$\tilde{T}^{(s)}_{nm}(\omega_i, \omega_j; 2X) = \sum_{l=m}^{n} \sum_{k=1}^{G} C_k \tilde{T}^{(s)}_{nl}(\omega_i, \omega_k; X) \tilde{T}^{(s)}_{lm}(\omega_k, \omega_j; X) + \left[\exp\left(-\frac{\Sigma_n}{\omega_i} X\right) + \exp\left(-\frac{\Sigma_m}{\omega_j} X\right)\right] \tilde{T}^{(s)}_{nm}(\omega_i, \omega_j; X). \tag{4.36}$$

By successive applications of the above functional relation we can readily obtain the function for thicknesses $2h$, $4h$, ..., $2^n h$, ... from the solution for the initial thickness h.

Thus, one can obtain solutions for very thick slabs without difficulty, even though one starts from the solution for a very thin slab.

From Eq. (4.21) we also have

$$\tilde{T}^{(s)}_{nm}(\omega_i, \omega_j; nX + X) = \sum_{l=m}^{n} \sum_{k=1}^{G} C_k \tilde{T}^{(s)}_{nl}(\omega_i, \omega_k; X) \tilde{T}^{(s)}_{lm}(\omega_k, \omega_j; nX) + \exp\left(-\frac{\Sigma_n}{\omega_i} X\right) \tilde{T}^{(s)}_{nm}(\omega_i, \omega_j; nX) + \tilde{T}^{(s)}_{nm}(\omega_i, \omega_j; X) \exp\left(-\frac{\Sigma_m}{\omega_j} nX\right). \tag{4.37}$$

Equation (4.37) can be used to generate solutions for thicknesses $2X$, $3X$, ..., nX, ... from the solution for a thickness X.

Equation (4.37) can be reduced further to a simpler form for the case when transmission due to an incident current of a specific energy spectrum and angular distribution is desired.

Let the current density of incident photons be $J(E, \omega; 0)$ and the corresponding current density with the multigroup approximation be $J_n(\omega, 0)$, where

$$J_n(\omega, 0) = \int_n dE\, J(E, \omega; 0). \tag{4.38}$$

We introduce the modified transmission current for scattered photons defined by the equation

$$\tilde{J}^{(s)}_n(\omega_i, X) = \sum_{m=1}^{n} \sum_{j=1}^{G} C_j \tilde{T}^{(s)}_{nm}(\omega_i, \omega_j; X) J_m(\omega_j; 0). \tag{4.39}$$

Then, from Eq. (4.37), we have

$$\tilde{J}^{(s)}_n(\omega_i, nX + X) = \sum_{m=1}^{n} \sum_{j=1}^{G} C_j \tilde{T}^{(s)}_{nm}(\omega_i, \omega_j; X) \times \left[\tilde{J}^{(s)}_m(\omega_j, nX) + \exp\left(-\frac{\Sigma_m}{\omega_j} nX\right) J_m(\omega_j; 0)\right] + \exp\left(-\frac{\Sigma_n}{\omega_i} X\right) \tilde{J}^{(s)}_n(\omega_i, nX). \tag{4.40}$$

Equation (4.40) involves the product of a matrix and a vector, which is less laborious than the product of two matrices involved in Eq. (4.37).

4.3 Accuracy of Numerical Solutions

4.3.1 Spatial Integration

The modified transmission function obtained by solving Eq. (4.35) for an initial thickness may contain an error. The error will accumulate with each successive application of the functional relation (4.36). This cumulative error, however, can be reduced to a negligible amount without difficulty. Let the relative error involved in the function for an initial thickness h be δ. We assume that

$$\delta \propto h^m. \tag{4.41}$$

The exponent m in Eq. (4.41) is at least 4, when the Runge–Kutta method is used to solve Eq. (4.35). Then, the cumulative error of the function for thickness $2^n h$ obtained by repeated applications of Eq. (4.36) is estimated to be $2^n \cdot \delta$ or less.

Suppose we reduce the initial thickness to $h/2$. Then, the corresponding error of the function for thickness $2^n \cdot h$ becomes $2^{n+1-m} \cdot \delta$. Thus, the cumulative error can be reduced by at least a factor of 2^{m-1} (= 8 when the Runge–Kutta method is used) by reducing the initial thickness to one-half. It should be emphasized here that it is important to use the Runge–Kutta method or other accurate methods to obtain the solution for initial thickness. At first, Eq. (4.18) seems solved for a very small thickness without using a laborious method such as the Runge–Kutta method. In fact, we can obtain an analytical solution by neglecting photons scattered more than once. Such a solution is accurate for thin slabs. Yet, it is not accurate enough to be used to generate the functions for thicker slabs, since the relative error involved in the solution is proportional to the thickness h [$m = 1$ in Eq. (4.41)].

The actual calculations for gamma rays have been made using an initial thickness of 1/64 mean free path at 1 MeV. The accuracy of these solutions was checked by comparison with solutions computed using an initial thickness of 1/128 mean free path. The discrepancy between both solutions was less than 10^{-2} at the thickness of 32 mean free paths. Even though one starts from an initial thickness as small as 1/64 mean free path, one can reach the thickness of 32 mean free paths by applying the functional relation (4.36) only 12 times.

4.3.2 Multigroup Approximation

It follows from the multigroup approximation that photons within a group interact with matter with a constant total cross section, even though the actual cross section varies with energy even within a group. Therefore, the widths of energy groups must be sufficiently small so that the variation of the cross section within a group may not have any effect on the photon reflection and transmission.

Transmission through thick slabs is controlled by photons whose energies are close to the energy corresponding to the smallest value of the cross section in the energy range below the source energy. Therefore, to reach a large thickness, the widths of energy groups around the energy corresponding to the smallest value of the total cross section should be chosen sufficiently small. Some numerical trial calculations indicate that the transmission functions obtained with a multigroup set are reliable for thicknesses less than a limit given roughly by the equation

$$X \simeq \frac{C}{\Sigma_{n+1} - \Sigma_n}, \tag{4.42}$$

where Σ_n is the total cross section of the group including the energy corresponding to the smallest value of the total cross section and Σ_{n+1} is that of the next lower group, and

$$C = 1.5 \sim 2.5 \qquad \text{for nearly normal incidence} \quad (\omega_0 \simeq 1),$$

$$C = 0.5 \sim 1.0 \qquad \text{for very oblique incidence} \quad (\omega_0 \simeq 0).$$

The multigroup sets used in a series of calculations of the transmission of gamma rays are shown in Tables 4.1–4.3. It is estimated that the total number of transmitted photons obtained with these multigroup sets and 7 angular divisions is accurate to within 10% error up to the thickness given by Eq. (4.42).

On the other hand, the reflection function is rather sensitive to the widths of energy groups at lower energies where the total cross section changes rapidly. The dependence of the reflection functions on the widths of energy groups was examined by comparing the results of a set which had 8 groups in the range 0.05 to 2.0 MeV with the results of another set that consisted of 13 groups in the same energy range. The relative

Table 4.1 *Multigroup Set I,* $E_0 \leq 1.0$ *MeV*

Group	Energy range (MeV)	Average energy (MeV)	Group width (CWL)[a]	$\Sigma_{n+1}/\Sigma_n - 1$	
				Water	Lead
1	1.05–0.95	1.00	0.051	0.054	0.10
2	0.95–0.85	0.90	0.063	0.049	0.113
3	0.85–0.77	0.81	0.062	0.046	0.119
4	0.77–0.69	0.73	0.077	0.047	0.124
5	0.69–0.63	0.66	0.071	0.046	0.130
6	0.63–0.57	0.60	0.085		
7	0.57–0.47	0.52	0.191		
8	0.47–0.35	0.41	0.373		
9	0.35–0.25	0.30	0.584		
10	0.25–0.17	0.21	0.961		
11	0.17–0.12	0.145	1.253		
12	0.12–0.09	0.105	1.419		
13	0.09–0.07	0.08	1.622		
14	0.07–0.05	0.06	2.92		

[a] CWL = Compton wavelength.

Table 4.2 *Multigroup Set II,* $E_0 \leq 3.0$ *MeV*

Group	Energy range (MeV)	Average energy (MeV)	Group width (CWL)	$\Sigma_{n+1}/\Sigma_n - 1$	
				Water	Lead
1	3.30–2.70	3.00	0.0345	0.106	0.029
2	2.70–2.30	2.50	0.0329	0.115	0.047
3	2.30–1.90	2.10	0.0467	0.089	0.064
4	1.90–1.60	1.75	0.0505	0.091	0.073
5	1.60–1.40	1.50	0.0456	0.078	0.086
6	1.40–1.20	1.30	0.0608		
7	1.20–1.00	1.10	0.0852		
8	1.00–0.80	0.90	0.1278		
9	0.80–0.60	0.70	0.2129		
10	0.60–0.40	0.50	0.4285		
11	0.40–0.24	0.32	0.8517		
12	0.24–0.14	0.19	1.521		
13	0.14–0.10	0.12	1.46		
14	0.10–0.08	0.09	1.37		

Table 4.3 *Multigroup Set III, $E_0 \leq 10.0$ MeV*

Group	Energy range (MeV)	Average energy (MeV)	Group width (CWL)	$\Sigma_{n+1}/\Sigma_n - 1$	
				Water	Lead
1	11.00–9.00	10.0	0.0103	0.092	0.059
2	9.00–7.20	8.1	0.0142	0.103	0.056
3	7.20–5.80	6.5	0.0171	0.100	0.038
4	5.80–4.80	5.3	0.0184	0.102	0.023
5	4.80–4.00	4.4	0.0213	0.103	0.013
6	4.00–3.30	3.65	0.0270		
7	3.30–2.50	2.90	0.0496		
8	2.50–1.70	2.10	0.0962		
9	1.70–1.10	1.40	0.1639		
10	1.10–0.64	0.87	0.3339		
11	0.64–0.36	0.50	0.6206		
12	0.36–0.22	0.29	0.9033		
13	0.22–0.14	0.18	1.327		
14	0.14–0.10	0.12	1.460		

deviation of the total number of reflected photons obtained with the 8-group set with that computed by the 13-group set is shown in Table 4.4. The deviation is less than about 5%. Therefore, it is estimated that the accuracy of the 13-group set itself will be enough to practical calculations.

Table 4.4 *Deviation of 8-Group Calculation from 13-Group Calculation*

E_0 (MeV)	Number albedo (%)				Energy albedo (%)			
	Water	Concrete	Iron	Lead	Water	Concrete	Iron	Lead
0.2	0.2	0.6	0.5	1.2	3.0	3.3	3.2	3.0
0.1	0.3	0.2	0.6	—	1.5	1.8	0.8	—
0.05	3.8	3.5	5.8	—	2.8	3.7	4.2	—

4.3.3 Gaussian Approximation

Gaussian approximation is somewhat similar to an artificial quantization of the continuous angular variable. Since changes in direction are restricted to discrete values, there is a minimum angular deflection for this approximation, even though the actual deflection is arbitrarily small. Therefore, the number of angular divisions required for numerical solutions depends on the smallest angle of deflection being considered. Since the deflection angle in the scattering correlates with the variation of energy, the smallest deflection angle to be considered is the function of the widths of energy groups in the multigroup set adopted. In Tables 4.1–4.3 the widths of energy groups are shown in Compton wavelength units.

Past experience, together with some numerical trial calculations, indicates the following rule:

Five divisions in the range $0 \leq \omega \leq 1$ are required when the width of the energy group in Compton wavelength units is more than 0.1, and 7 divisions are necessary for the widths between 0.1 and 0.01 Compton wavelength.

CHAPTER FIVE *REFLECTION AND TRANSMISSION OF GAMMA RAYS*

A series of calculations was made on the reflection and transmission of gamma rays by the method of invariant embedding by Shimizu *et al.* [22–27]. The method of numerical solutions described in the previous chapter was used in the calculations. The results of the calculations will be given here.

5.1 Reflection by Semi-Infinite Media

5.1.1 Albedo

The reflecting power of a bounded medium for gamma rays is customarily expressed by the quantity called "albedo." In terms of the reflection function the albedo is expressed by the following equations:

number albedo

$$R(E_0, \omega_0; X) = \int_0^{E_0} dE \int_0^1 d\omega\, R(E, \omega \mid E_0, \omega_0; X), \tag{5.1}$$

energy albedo

$$R_E(E_0, \omega_0; X) = \int_0^{E_0} dE \int_0^1 d\omega\, \frac{E}{E_0} R(E, \omega \mid E_0, \omega_0; X), \tag{5.2}$$

Table 5.1 *Comparison with Other Method for Albedos*[a]

No.	Quantities Compared	Comparison method	Authors	Source energy (MeV)	Type of source	Medium	Thickness (mfp)	Result of comparison
1	Number and energy albedo	Monte Carlo	Berger and Raso [32]	0.02–1	Plane oblique Plane isotropic	H_2O, Fe, Sn, W, Pb, concrete	Semi-infinite medium	Maximum discrepancy: 10% for Pb 5% for others
2	Energy spectrum and angular distribution of backscattered photons	Monte Carlo	Berger and Raso [31]	1	Plane isotropic	H_2O, Fe, Sn, W, Pb, concrete	Semi-infinite medium	Good agreement

[a] Shimizu [24].

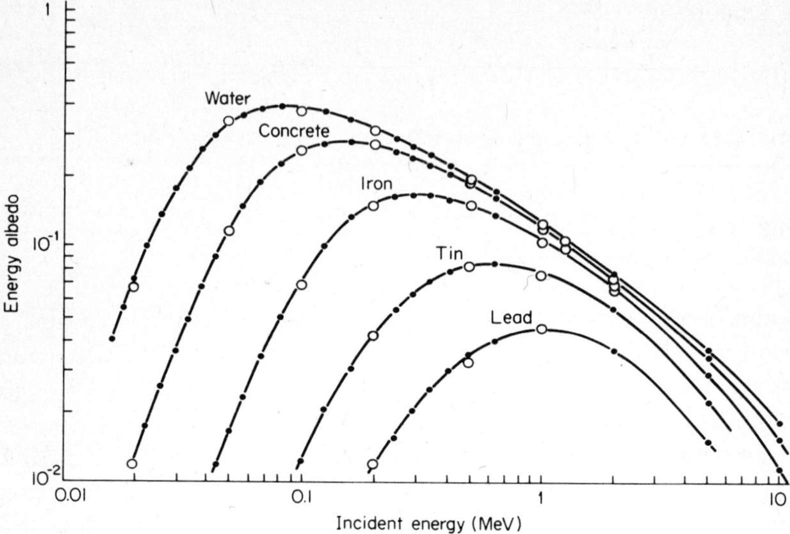

Fig. 5.1 *Energy albedo of isotropic incident gamma rays on various substances.* ○ *Monte Carlo calculation by Berger and Raso* [32], ● *invariant embedding calculation* [22].

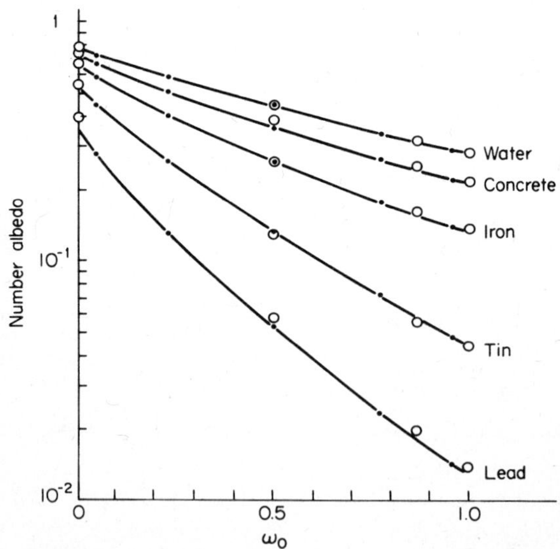

Fig. 5.2 *Number albedo of obliquely incident gamma rays on various substances.* ○ *Monte Carlo calculation by Berger and Raso* [32], ● *invariant embedding calculation* [22]. *The abscissa is the cosine of the incident obliquity with respect to the normal of the surface.*

dose albedo

$$R_D(E_0, \omega_0; X) = \int_0^{E_0} dE \int_0^1 d\omega \frac{\Sigma_a(E)E}{\Sigma_a(E_0)E_0} R(E, \omega \mid E_0, \omega_0; X), \tag{5.3}$$

where $\Sigma_a(E)$ is the energy absorption coefficient of air.

5.1.2 Comparison with Monte Carlo Calculations

A number of calculations was made to compare the results of invariant embedding calculations with those obtained by the Monte Carlo method. Most calculations were carried out with 14 energy groups and 5 angular divisions.

The results of the comparisons are summarized in Table 5.1. The energy albedos for isotropic sources as functions of the source energy computed both by the method of invariant embedding and the Monte Carlo method (comparison calculation No. 1) are plotted in Fig. 5.1.

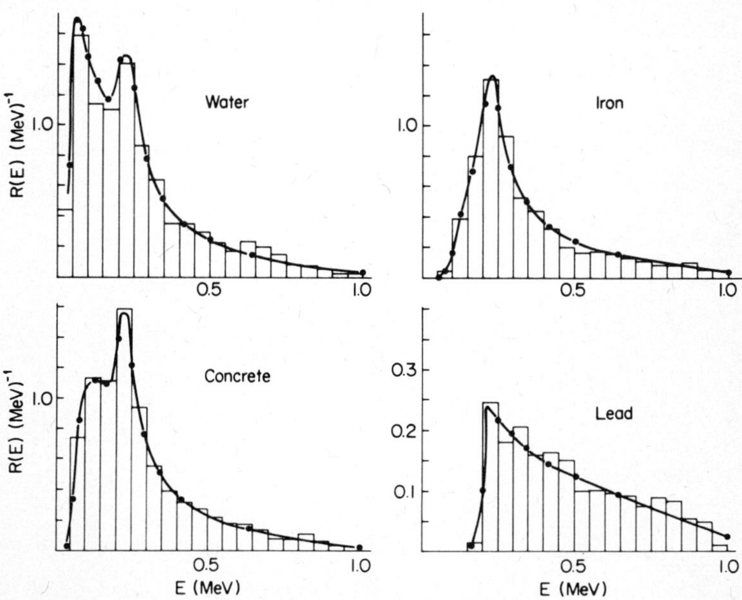

Fig. 5.3 *Energy spectra of reflected gamma rays from various substances. Incident current is isotropic with energy* 1 *MeV. Bars indicate Monte Carlo calculation by Berger and Raso* [31], *curve indicates invariant embedding calculation* [22].

The Monte Carlo calculations were carried out by Berger and Raso [32]. The maximum discrepancy between both calculations is about 10% for lead and 5% for other elements.

The number albedos of obliquely incident radiations are shown in Fig. 5.2 as functions of the cosine of the source obliquity.

The energy spectra and the angular distributions of photons back-scattered from semi-infinite media computed by the method of invariant embedding are compared with the results of Monte Carlo calculations made by Berger and Raso [31] (comparison calculation No. 2) in Figs. 5.3 and 5.4.

The figures indicate that the agreement between the invariant embedding calculations and the Monte Carlo ones is satisfactory.

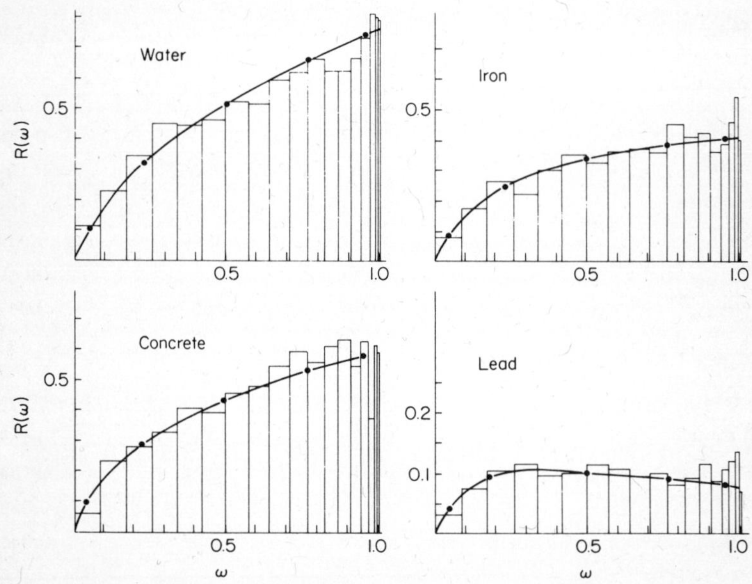

Fig. 5.4 *Angular distributions of reflected gamma rays from various substances. Incident current is isotropic with energy* 1 *MeV. Bars indicate Monte Carlo calculation by Berger and Raso* [31], *curve indicates invariant embedding calculation* [22].

5.1.3 Comparison with Experiment

The energy albedos of various substances for collimated gamma rays from ^{60}Co and ^{198}Au have been measured by Bulatov and Garusov [29]. The number and energy albedos of gamma rays from isotropic point sources of ^{60}Co and ^{137}Cs have been measured by Bulatov [33] and Hyodo [34]. Hyodo [34] has also obtained the energy spectra and angular distributions of the reflected photons using a scintillation spectrometer.

The energy albedos for isotropic sources measured by Bulatov and Hyodo are compared with the invariant embedding calculations in Table 5.2. The albedo obtained by Hyodo is larger, and Bulatov's smaller, than the calculated value. The discrepancy between Hyodo's values and theory lies between 10 and 30%.

Table 5.2 *Energy Albedo of Gamma Rays from a Point Isotropic Source of ^{60}Co and ^{137}Cs*[a]

Source	Scatterer	Measurement		Calculation
		Hyodo	Bulatov	
Paraffin	^{60}Co	0.112	—	0.108
Graphite		—	0.091	0.109
Aluminum		0.136	0.094	0.104
Iron		0.120	0.077	0.093
Tin		0.088	—	0.068
Lead		0.059	0.029	0.043
Paraffin	^{137}Cs	0.19	—	0.168
Graphite		—	0.12	0.172
Aluminum		0.20	0.13	0.159
Iron		0.16	0.16	0.134
Tin		0.10	—	0.085
Lead		0.04	0.04	0.041

[a] Measurements were made by Hyodo [34] and Bulatov [33] and the calculation by Shimizu and Mizuta [22].

The energy albedo for oblique incident photons obtained both by experiment [29] and theory [22] is plotted as a function of the atomic number of the substance in Fig. 5.5. Agreement between theory and experiment is fair.

The energy spectra of gamma rays backscattered from paraffin, aluminum, iron, tin, and lead measured by Hyodo are compared with cal-

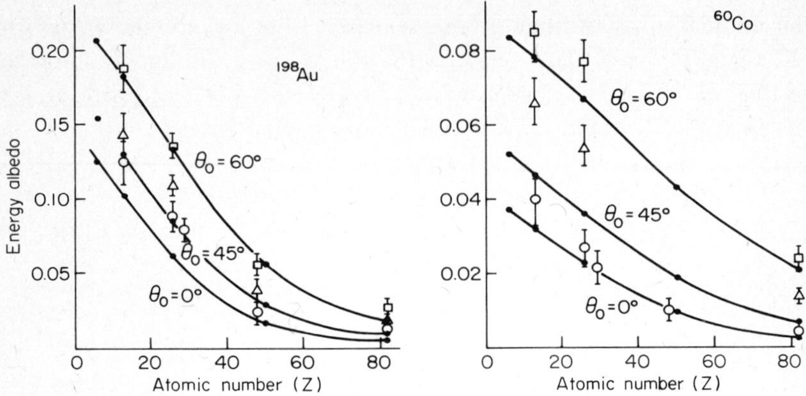

Fig. 5.5 *Energy albedo of collimated gamma rays.* □ ($\theta_0=60°$), △ ($\theta_0=45°$), ○ ($\theta_0=0°$), *measurement by Bulatov and Garusov* [29]; ● *invariant embedding calculation* [22].

culations in Fig. 5.6. The energy spectrum plotted in the figure is a normalized one given by

$$\int_0^1 d\omega\, R(E, \omega \mid E_0, \omega_0; \infty) \Big/ \int_0^{E_0} dE \int_0^1 d\omega\, R(E, \omega \mid E_0, \omega_0; \infty).$$

Agreement between theory and experiment is remarkable in the energy range above 0.3 MeV.

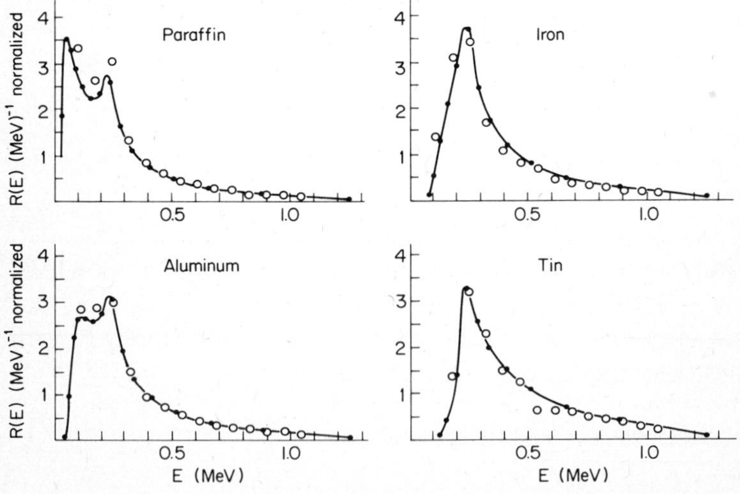

Fig. 5.6 *Energy spectra of reflected gamma rays. Source: isotropic point source of* ^{60}*Co.* ○ *measurement by Hyodo* [34], ● *invariant embedding calculation* [22].

The angular distributions of the reflected photons are shown in Fig. 5.7. For iron and tin, the experimental points are well aligned along the theoretical curve. A large discrepancy, however, is observed for paraffin. The reason for discrepancy has not yet been clarified.

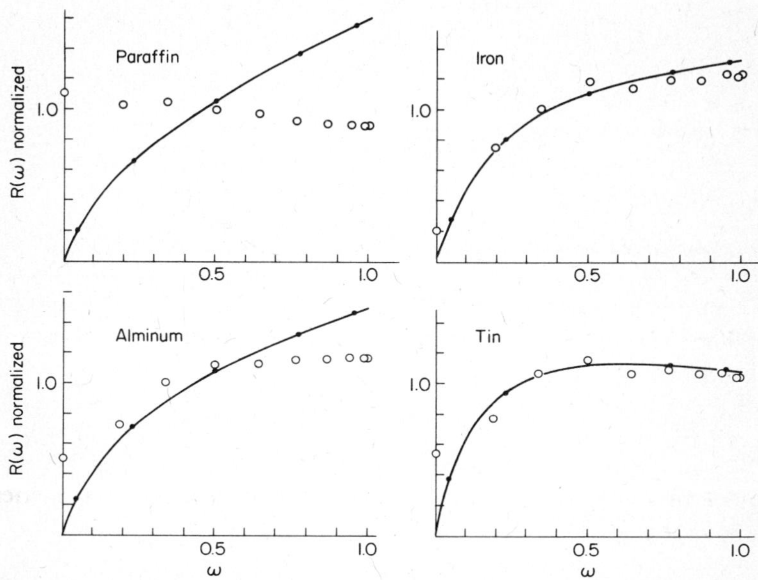

Fig. 5.7 *Angular distribution of reflected gamma rays. Source: Isotropic point source of* ^{60}*Co.* ○ *measurement by Hyodo* [34], ● *invariant embedding calculation* [22].

5.1.4 Fraction of Singly Scattered Photons

A distinction may be made between the reflected photons which have had only one collision in the medium and those which have suffered collisions more than once. The reflection function can be written as the sum of two components:

$$R(E, \omega \mid E_0, \omega_0; \infty) = R^{(1)}(E, \omega \mid E_0, \omega_0; \infty) + R'(E, \omega \mid E_0, \omega_0; \infty). \tag{5.4}$$

The first term represents the contribution of singly scattered photons and the second term that of multiply scattered ones. The reflection function for singly scattered photons can be expressed analytically by the equation

$$R^{(1)}(E, \omega \mid E_0, \omega_0; \infty) = \frac{\Sigma_s(E_0, \omega_0 \to E, -\omega)}{\omega_0[\Sigma(E)/\omega + \Sigma(E_0)/\omega_0]}. \tag{5.5}$$

Equation (5.5) is derived from Eq. (2.35) by setting the reflection function on the right-hand side to zero.

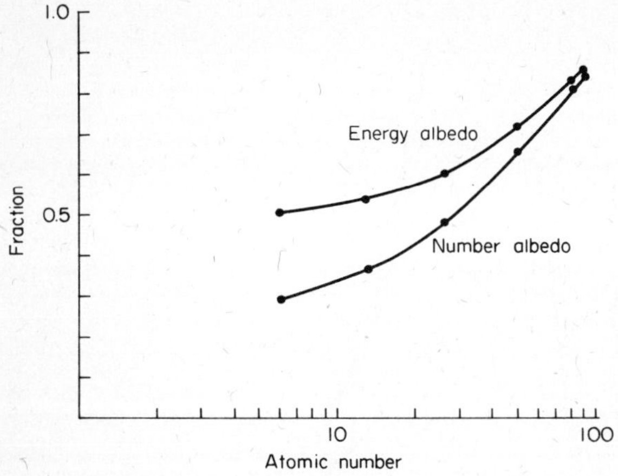

Fig. 5.8 *Fraction of singly scattered photons in albedo.* $E_0 = 1.0$ *MeV*, ω_0 *isotropic.*

The fraction of singly scattered photons in the albedo was computed by numerical solutions. It increases with the atomic number of the substance as shown in Fig. 5.8. The fraction in the number albedo for graphite is 30%, whereas it amounts to 80% for lead.

5.1.5 Energy Spectra

The energy spectra of photons backscattered from various substances are shown in Fig. 5.9. The spectrum has a peak at energy between 0.2 and 0.25 MeV irrespective of scatterers. The location of the peak is independent of the obliquity of the incident photons as shown in Fig. 5.10 for iron. Spectra of singly and multiply scattered photons are plotted separately for graphite in Fig. 5.11. It is seen that the peak is formed by singly scattered photons.

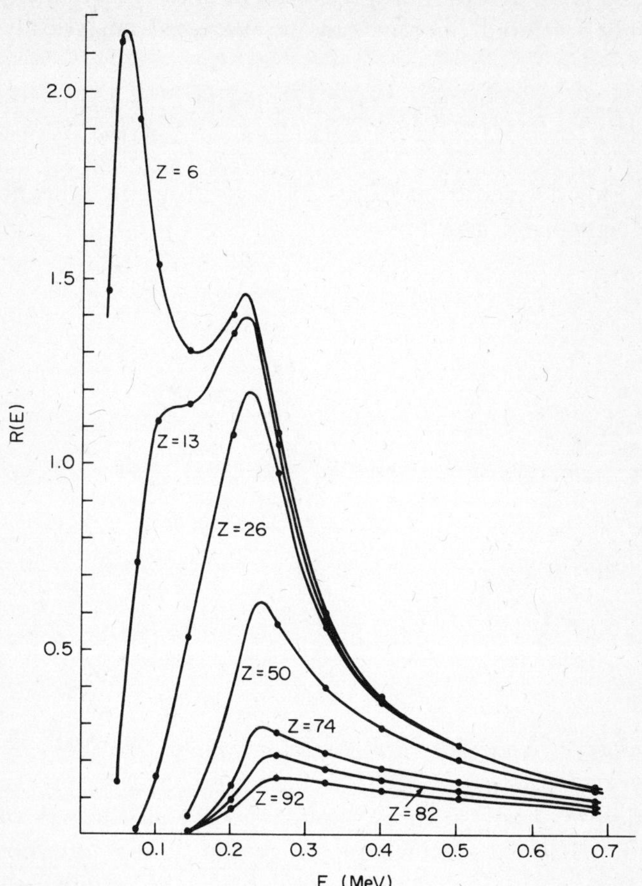

Fig. 5.9 *Energy spectra of reflected gamma rays from various substances.* $E_0 = 1.0$ *MeV,* ω_0 *isotropic.*

The spectra of photons backscattered from light elements have additional peaks or discontinuities at lower energy. Spectra of reflected photons in the case of oblique incidence are shown in Fig. 5.12. Two more peaks or discontinuities are observed for water. There might be additional peaks in the energy range below 0.08 MeV. The distance between peaks is about 2 Compton wavelength units. The location of those peaks appears to be independent of the scatterers as well as of source obliquity. The peak height, however, does change appreciably with the scatterers and also with source obliquity.

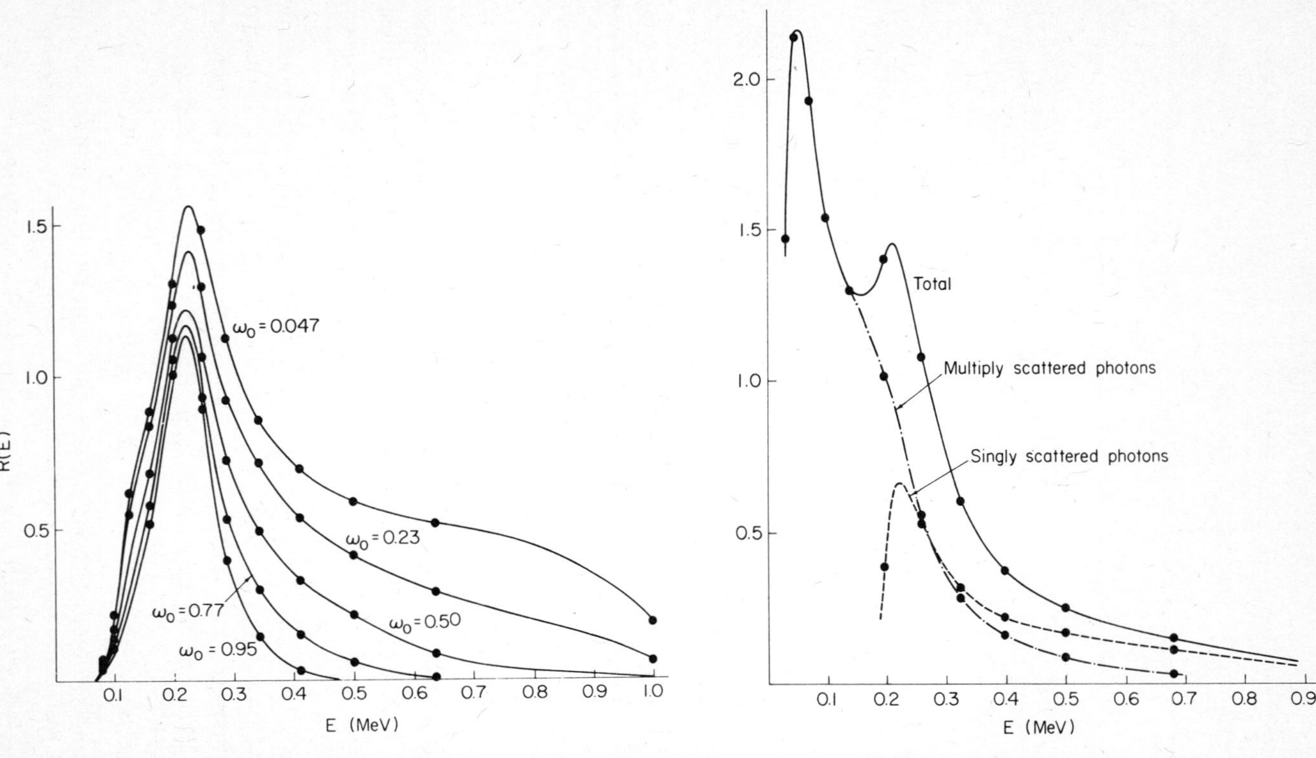

Fig. 5.10 *Energy spectra of reflected gamma rays from iron. $E_0 = 1.0$ MeV.*

Fig. 5.11 *Energy spectrum of reflected gamma rays from graphite. $E_0 = 1.0$ MeV, ω_0 isotropic.*

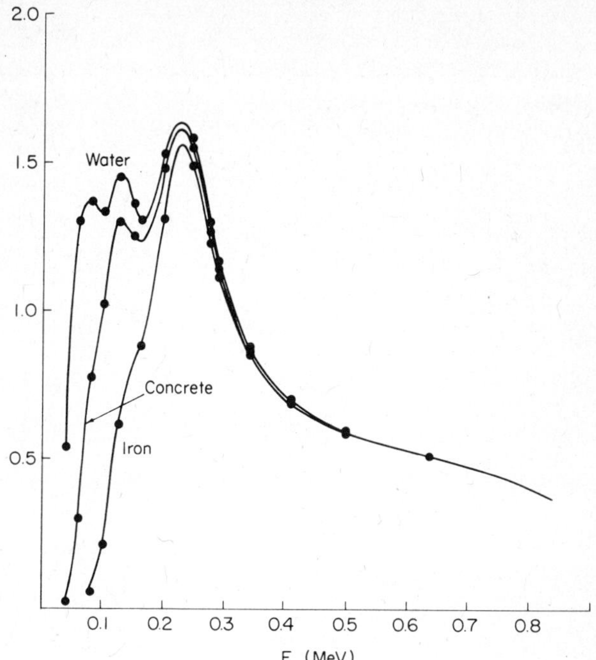

Fig. 5.12 *Energy spectra of reflected gamma rays from water, concrete, and iron.* $E_0 = 1.0$ *MeV,* $\omega_0 = 0.047$.

5.1.6 Angular Distributions

Angular distributions of photons backscattered from various substances due to isotropic incident photons are shown in Fig. 5.13. The intensity of photons backscattered into the direction perpendicular to the surface increases as the atomic number decreases.

Angular distributions of photons backscattered from water and lead due to oblique incident photons are shown in Fig. 5.14. Angular distributions of singly and multiply scattered photons are also plotted in the figure. It is seen that the angular distribution of multiply scattered photons from water is markedly peaked in the direction perpendicular to the surface.

We shall consider more closely the angular distributions of singly scattered photons to find the factors governing them. When photons are incident normally to the surface ($\omega_0 = 1$), the angular distribution of singly scattered photons is expressed as

$$R^{(1)}(\omega \mid E_0, 1) = \int_0^{E_0} dE\, R^{(1)}(E, \omega \mid E_0, 1; \infty)$$
$$= \frac{2\pi N_e \sigma(E_0, -\omega)}{\Sigma(E_0) + \Sigma(E)/\omega}, \tag{5.6}$$

where $\sigma(E_0, \cos\theta)$ is the Klein–Nishina differential cross section given by Eq. (1.3) and N_e is the electron density. The total cross section $\Sigma(E)$ in Eq. (5.6) should be considered as a function of ω by the relation

$$E = \frac{E_0}{1 + (E_0/mc^2)(1 + \omega)}, \tag{5.7}$$

where mc^2 is the electron rest mass (0.511 MeV).

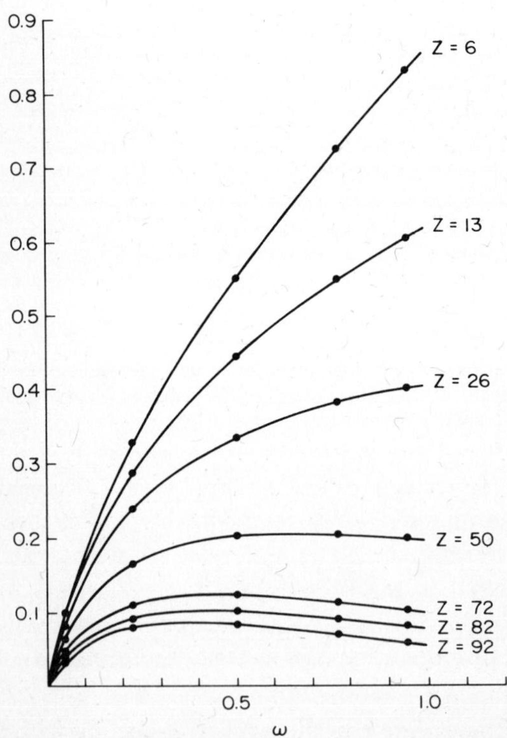

Fig. 5.13 *Angular distributions of reflected gamma rays from various substances.* $E_0 = 1.0$ *MeV,* ω_0 *isotropic.*

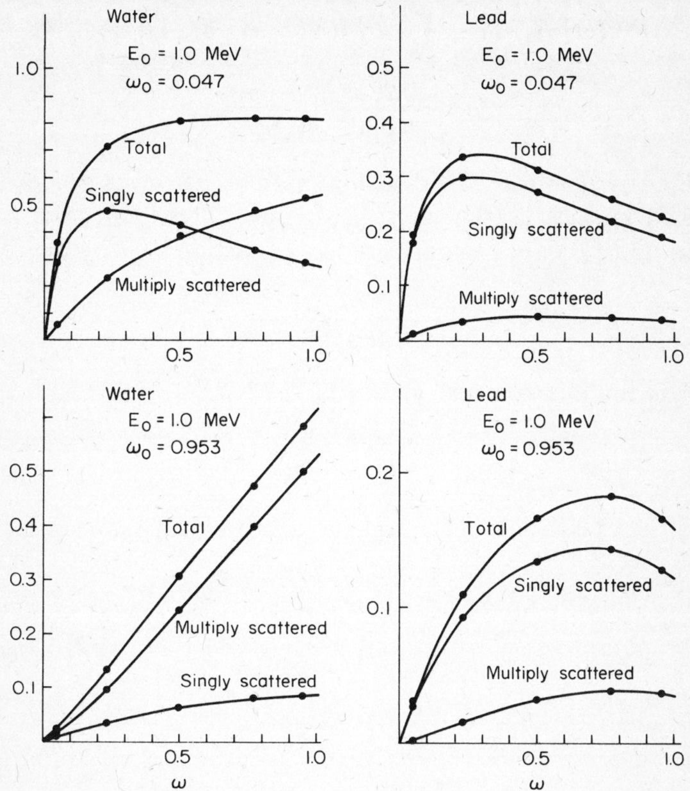

Fig. 5.14 *Angular distribution of reflected photons from water and lead.*

The numerator and the denominator on the right-hand side of Eq. (5.6) are plotted for water and lead in Fig. 5.15. It is seen that the differential cross section $\sigma(E_0, -\omega)$ is almost isotropic in this case, and that the angular distribution is governed mostly by the quantity $\Sigma(E)/\omega$ which is proportional to the probability that photons scattered into the backward direction ω will have a collision before reaching the free surface. The smaller the quantity $\Sigma(E)/\omega$, the more photons will reach the surface.

It is noted that the variation of the total cross section with energy around 0.2 or 0.1 MeV plays an important role in forming the angular distribution through Eq. (5.7). The trend in the angular distribution observed in Fig. 5.13 may be explained as follows. The total cross section of a substance with higher atomic number increases more rapidly than that of a low-Z material as energy decreases. Consequently, the curve of the quantity $\Sigma(E)/\omega$ as a function of ω becomes flatter as shown in Fig. 5.14. The angular distribution from the heavier element, therefore, is less peaked in the direction perpendicular to the surface.

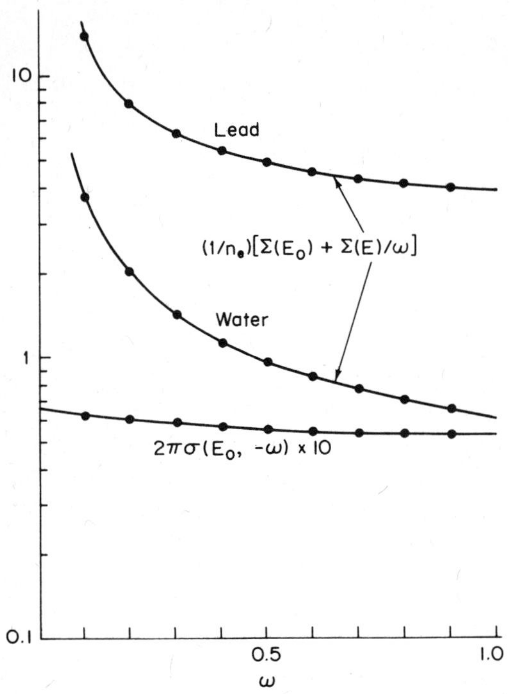

Fig. 5.15 *Angular distribution of singly scattered photons.* $E_0 = 1.0$ *MeV*, $\omega_0 = 1.0$.

5.2 Transmission through Homogeneous Slabs

By using the method of numerical solutions given in the previous chapter, it is easy to compute the modified transmission function for a homogeneous slab of finite thickness, once the reflection function for a semi-infinite medium is obtained. The ordinary transmission function can also be obtained by conversion from the modified transmission function. For slabs whose thickness is greater than 1 or 2 mean free paths at the source energy, the conversion from the modified transmission function to the ordinary one can be made by Eq. (2.45) which involves only the reflection function for a semi-infinite medium. The conversion for very thin slabs, however, requires the reflection function as the function of slab thickness, which, in principle, cannot be obtained unless we solve the basic equation (2.25). It is, however, to be noted that the term $\mathbf{R}(\infty)\mathbf{R}(X)$ in the conversion formula (2.44) is usually very small compared with unity. Therefore, a crude solution for $\mathbf{R}(X)$ will be sufficient for the conversion. The actual computations of the transmission functions were carried out with the approximation

$$R(E, \omega \mid E_0, \omega_0; X) \simeq R(E, \omega \mid E_0, \omega_0; \infty) \times \left(1 - \exp\left[-\left\{\frac{\Sigma(E_0)}{\omega_0} + \frac{\Sigma(E)}{\omega}\right\}X\right]\right). \tag{5.8}$$

This approximation is rather crude for obtaining the reflection function itself for thin slabs, yet accurate enough to make the conversion from the modified transmission function to the ordinary one. The approximation in Eq. (5.8) eliminates the problem of solving the basic equation (2.25), which would be very time consuming.

5.2.1 Buildup Factor

It is convenient in shielding designs to express the transmission of gamma rays in terms of the buildup factor which is defined as the ratio of the total flux of transmitted radiations to the flux of unscattered radiations. For monodirectional sources, the number, energy, and dose buildup factors are given in terms of the transmission functions by the following equations:

number buildup factor

$$B(E_0, \omega_0; X) = \omega_0 \exp\left(\frac{\Sigma(E_0)}{\omega_0} X\right) \int_0^{E_0} dE \int_0^1 \frac{d\omega}{\omega} T(E, \omega \mid E_0, \omega_0; X), \tag{5.9}$$

energy buildup factor

$$B_{\mathrm{E}}(E_0, \omega_0; X) = \omega_0 \exp\left(\frac{\Sigma(E_0)}{\omega_0} X\right) \times \int_0^{E_0} dE \int_0^1 \frac{d\omega}{\omega} \frac{E}{E_0} T(E, \omega \mid E_0, \omega_0; X), \tag{5.10}$$

dose buildup factor

$$B_{\mathrm{r}}(E_0, \omega_0; X) = \omega_0 \exp\left(\frac{\Sigma(E_0)}{\omega_0} X\right) \times \int_0^{E_0} dE \int_0^1 \frac{d\omega}{\omega} \frac{E\Sigma_{\mathrm{a}}(E)}{E_0\Sigma_{\mathrm{a}}(E_0)} T(E, \omega \mid E_0, \omega_0; X), \tag{5.11}$$

where $\Sigma_{\mathrm{a}}(E)$ is the energy absorption coefficient of air.

For very oblique sources, the buildup factor becomes exceedingly large because of the exponential factor $\exp[\Sigma(E_0)X/\omega_0]$, although the number of transmitted radiations is small. Therefore, it is more convenient for oblique sources to introduce another factor, which we shall call the transmission factor. The transmission factor is defined as the ratio of the total flux of scattered radiations to the flux of unscattered radiations that are incident normally to the slab (the radiations from a plane perpendicular source). The number, energy, and dose transmission factors are expressed by the equations:

number transmission factor

$$F(E_0, \omega_0; X) = e^{\Sigma(E_0)X} \int_0^{E_0} dE \int_0^1 \frac{d\omega}{\omega} T^{(s)}(E, \omega \mid E_0, \omega_0; X), \tag{5.12}$$

energy transmission factor

$$F_{\mathrm{E}}(E_0, \omega_0; X) = e^{\Sigma(E_0)X} \int_0^{E_0} dE \int_0^1 \frac{d\omega}{\omega} \frac{E}{E_0} T^{(s)}(E, \omega \mid E_0, \omega_0; X), \tag{5.13}$$

dose transmission factor

$$F_{\mathrm{r}}(E_0, \omega_0; X) = e^{\Sigma(E_0)X} \int_0^{E_0} dE \int_0^1 \frac{d\omega}{\omega} \frac{\Sigma_{\mathrm{a}}(E)E}{\Sigma_{\mathrm{a}}(E_0)E_0} T^{(s)}(E, \omega \mid E_0, \omega_0; X), \tag{5.14}$$

Table 5.3 *Comparison with Other Method for Transmissions*[a]

1	Energy buildup factor	Moments method	Goldstein and Wilkins [7]	1 MeV		H_2O, Fe, Pb	Infinite medium; distance from source 1 to 15	Maximum discrepancy 10%
				10 MeV	PLP[b]	Fe, Pb		10%
				8 MeV		H_2O		20% at 15 mfp
2	Differential energy flux	Moments method	Goldstein and Wilkins [7]	1 MeV		H_2O, Fe, Pb	Infinite medium; distance from source 1, 2, 4	Good agreement except that the present solution for water is somewhat lower at low energy
3	Dose transmission and its angular distribution	Monte Carlo method	Berger and Morris [35]	1.25 MeV	PLO[c]	Concrete, Fe	1, 2, 4	Discrepancy is less than 10% except for the case that $\omega_0 \simeq 0$
	Differential energy-angular distribution	Monte Carlo method	Berger and Morris [36]	0.66 MeV	PLP	Al	1 to 4	Good agreement
4	Differential energy-angular distribution	Monte Carlo method	Berger and Morris [36]	10 MeV	PLP	Al	2, 3	Fair agreement
5	Dose transmission	Moments method	Spencer and Lamkin [37]	0.256, 1.0, 10.22 MeV	PLO	H_2O	Infinite medium; distance from source 0.5 to 10	Discrepancy is less than 10% except for the case that $\omega_0 \simeq 0$ at the distance above 4 mfp
6	Energy absorption buildup factor	Moments method	Berger [38]	0.255, 0.5, 1.0 MeV	PLI[d]	H_2O	Infinite medium; distance from source 0.5 to 16	Maximum discrepancy 5%

[a] Shimizu [24]. [b] PLP, plane perpendicular source. [c] PLO, plane oblique source. [d] PLI, plane isotropic source.

where $T^{(s)}(E, \omega \mid E_0, \omega_0; X)$ is the transmission function for scattered photons.

The dose transmission factors for various substances have been computed by the method of invariant embedding and tabulated as functions of the source energy, the cosine of the source obliquity, and the slab thickness [26].

5.2.2 Comparison with other Calculations

The invariant embedding calculations were compared with other calculations made by the method of moments and the Monte Carlo method. The results of the comparisons are summarized in Table 5.3. Most calculations by the method of invariant embedding were carried out with 7 angular divisions and 14 energy groups.

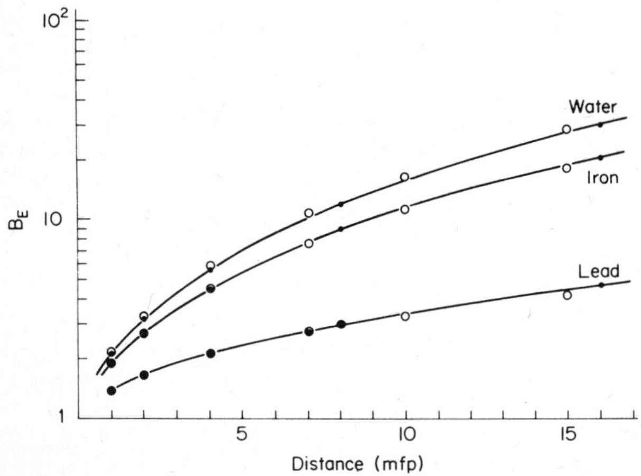

Fig. 5.16 *Energy buildup factor in an infinite medium. Substances: water, iron, and lead. Source: plane monodirectional, $E_0 = 1.0$ MeV.* ○ *moments method calculation by Goldstein and Wilkins* [7], ● *invariant embedding calculation* [23].

Figures 5.16–5.19 show the results of comparison with the moments method calculations made by Goldstein and Wilkins [7] (comparison calculations No. 1 and 2). Figures 5.16, 5.17, and 5.18 show the energy buildup factors in infinite media of water, iron, and lead for plane perpendicular sources. The differential energy flux in infinite media is

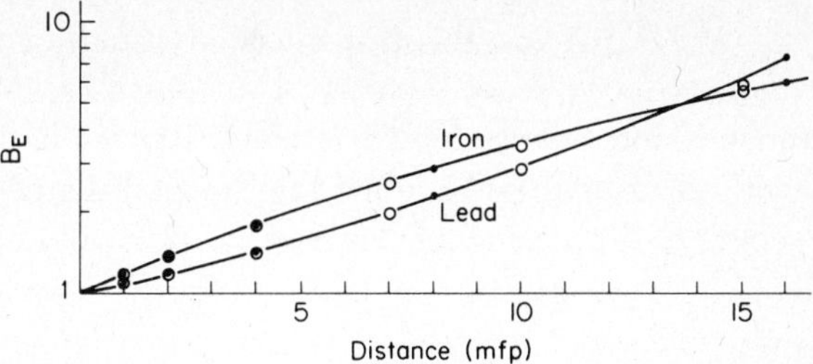

Fig. 5.17 *Energy buildup factor in an infinite medium. Substances: iron and lead. Source: plane monodirectional,* $E_0 = 10$ *MeV.* ○ *moments method calculation by Goldstein and Wilkins* [7], ● *invariant embedding calculation* [23].

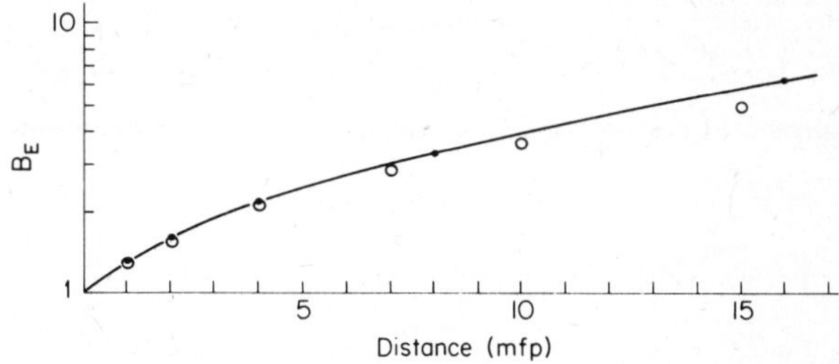

Fig. 5.18 *Energy buildup factor in an infinite medium. Substance: water. Source: plane monodirectional,* $E_0 = 8.0$ *MeV.* ○ *moments method calculation by Goldstein and Wilkins* [7], ● *invariant embedding calculation* [23].

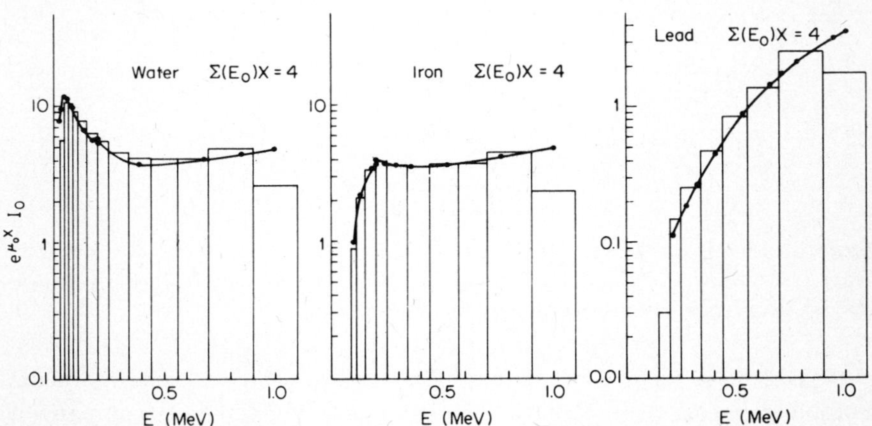

Fig. 5.19 *Differential energy flux in an infinite medium. Substances: water, iron, and lead. Source: plane monodirectional,* $E_0 = 1.0$ *MeV. Curves indicate moments method calculation by Goldstein and Wilkins* [7], *bars indicate invariant embedding calculation* [23].

plotted in Fig. 5.19. The discrepancy in the energy buildup factor between the moments method calculations and the invariant embedding ones is about 10% or less except for the buildup factor of water for an 8-MeV source, for which a discrepancy of about 20% is observed at the distance of 15 mean free paths. The reason for this discrepancy has been not clarified. It should, however, be mentioned that the problem of the penetration in water at energy around 10 MeV lies almost at the limit beyond which more than 7 angular divisions are required in the invariant embedding calculations.

The dose transmissions for concrete as functions of the cosine of the source obliquity computed both by the Monte Carlo method and the invariant embedding method are plotted in Fig. 5.20. Figure 5.21 shows the angular distributions of transmitted photons (comparison calculation No. 3).

The Monte Carlo calculations were carried out by Berger and Morris [35]. The figures indicate that the agreement between the invariant

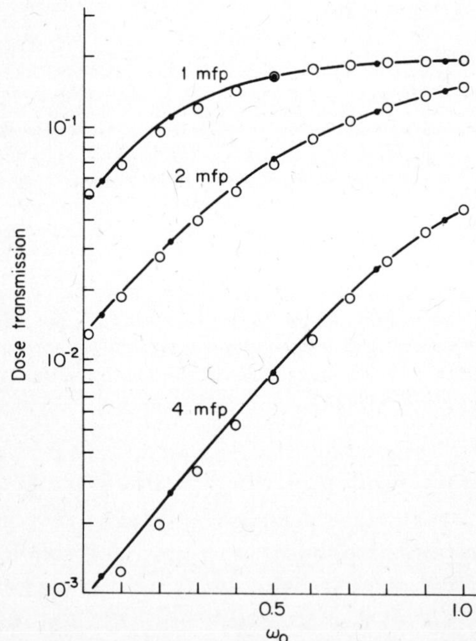

Fig. 5.20 *Dose transmission (current). Substance: concrete. Source: oblique,* $E_0 = 1.25$ *MeV.* ○ *Monte Carlo calculation by Berger and Morris* [35], ● *invariant embedding calculation* [23].

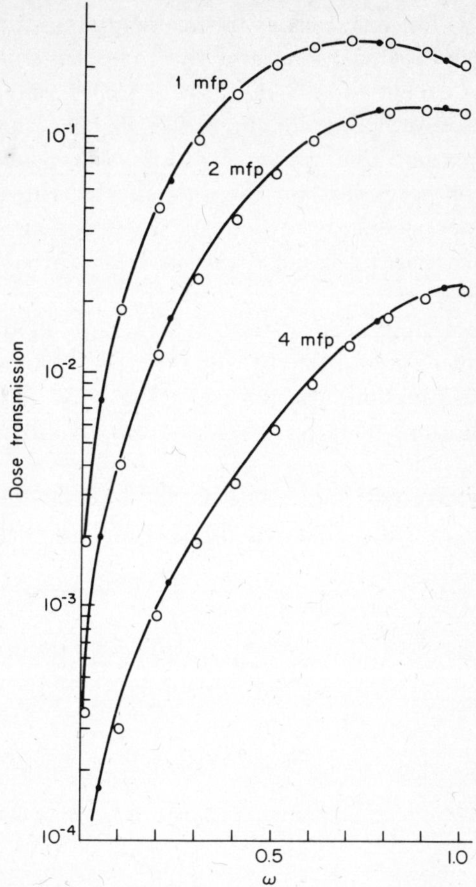

Fig. 5.21 *Angular distribution of dose transmission (current). Substance: concrete. Source:* $\omega_0 = 0.5$, $E_0 = 1.25$ *MeV.* ○ *Monte Carlo calculation by Berger and Morris* [35], ● *invariant embedding calculation* [23].

embedding calculations and the Monte Carlo ones is quite satisfactory.

The energy spectrum of photons transmitted through an aluminum slab at an angle of 60° from the perpendicular axis of the slab (comparison calculation No. 4) is plotted in Fig. 5.22. The solid line in the figure refers to the Monte Carlo calculation by Berger and Morris [36], and the dashed line to the invariant embedding calculation.

The results of comparison calculation No. 5 on oblique penetration of

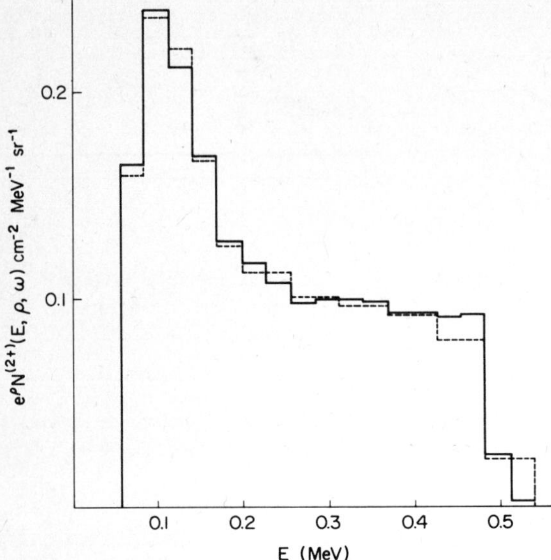

Fig. 5.22 *The energy spectrum of radiation of the second- and higher-order scattering from a* ^{137}Cs *plane perpendicular source transmitted through an aluminum slab of* 2.042 *mean free paths at an angle of* 60° *from the perpendicular axis of the slab.* — *Monte Carlo calculation by Berger and Morris* [36], - - - *invariant embedding calculation* [24].

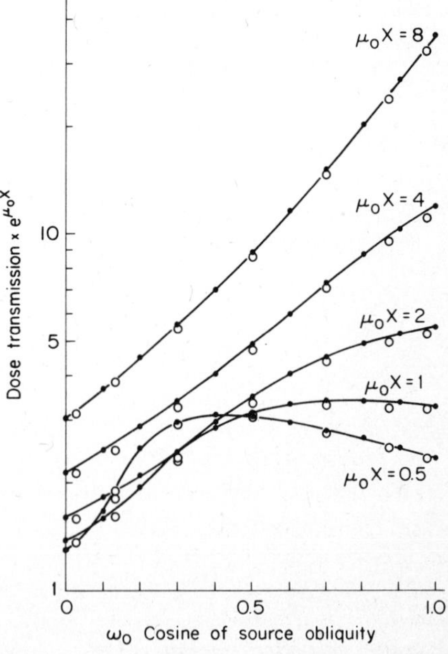

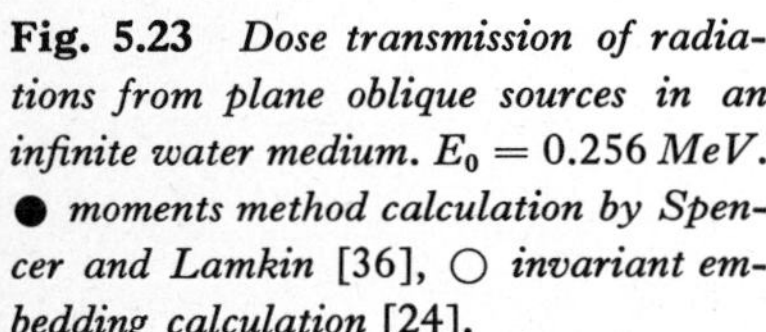
Fig. 5.23 *Dose transmission of radiations from plane oblique sources in an infinite water medium.* $E_0 = 0.256$ *MeV.* ● *moments method calculation by Spencer and Lamkin* [36], ○ *invariant embedding calculation* [24].

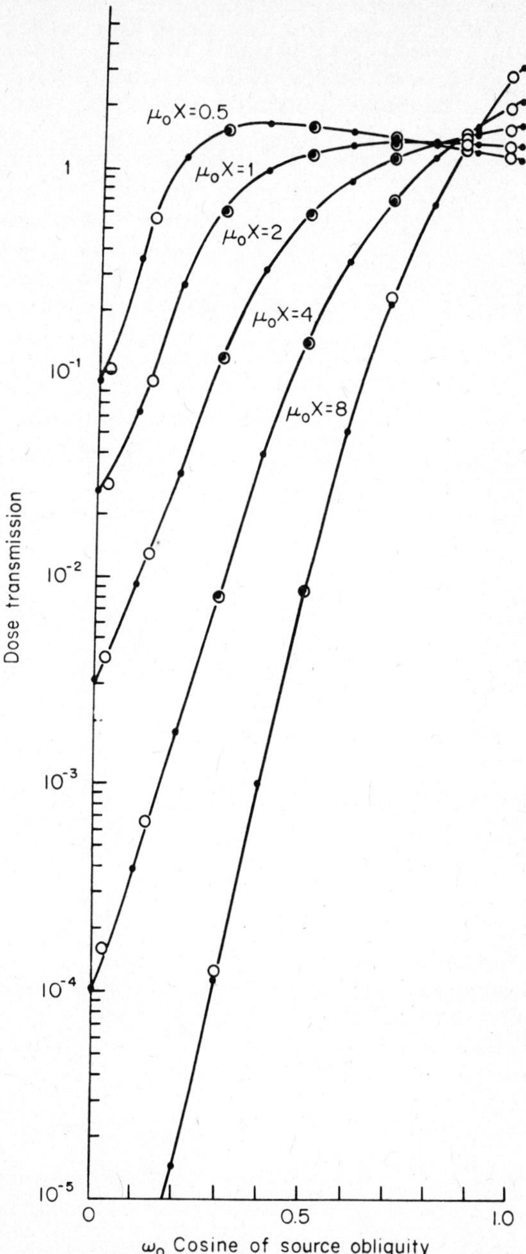

Fig. 5.24 *Dose transmission of radiations from plane oblique sources in an infinite water medium.* $E_0 = 10.22$ *MeV.* ● *moments method calculation by Spencer and Lamkin* [37], ○ *invariant embedding calculation* [24].

gamma rays in water are shown in Figs. 5.23 and 5.24, where the dose transmissions are plotted as functions of the cosine of the source obliquity. The discrepancy between the moments method calculations by Spencer and Lamkin [37] and the invariant embedding calculations is less than 10%.

5.2.3 Comparison with Experiments

The dose buildup factors of slabs were measured by Furuta *et al.* [39] for a ^{60}Co plane perpendicular source and by Tamura and Tsuruo [40]

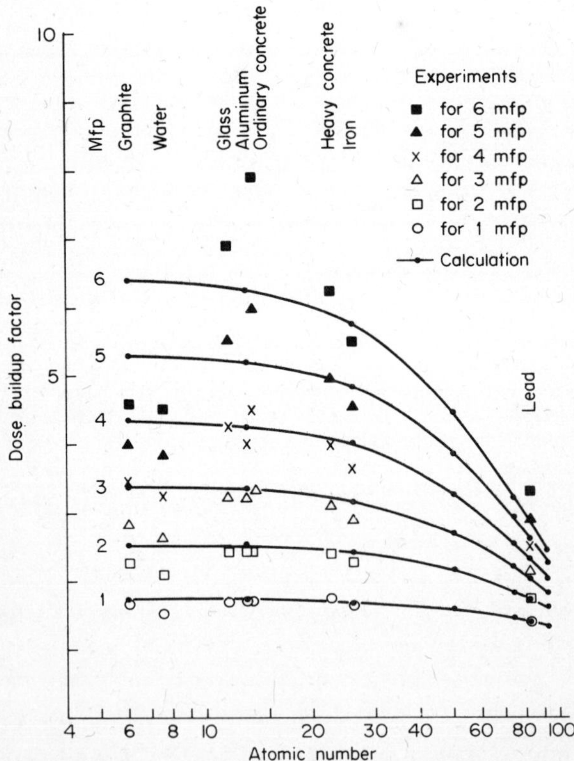

Fig. 5.25 *Dose buildup factors of radiations from a ^{60}Co plane perpendicular source. The measurements were made by Furuta et al.* [39] *and the calculations by Shimizu* [24].

for a ^{137}Cs plane perpendicular source. Figures 5.25 and 5.26 show the results of comparing the invariant embedding calculations with their measurements. Experimental and theoretical buildup factors agree within 20% up to 6 mean free paths for all the materials except water and graphite for a ^{60}Co source and water and lead for a ^{137}Cs source. The discrepancies for water and graphite are probably due to the measurements.

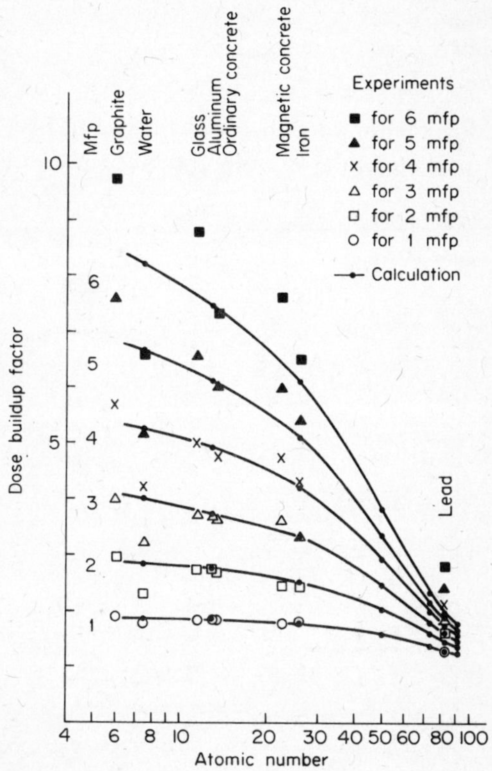

Fig. 5.26 *Dose buildup factors of radiations from a* ^{137}Cs *plane perpendicular source. The measurements were made by Tamura and Tsuruo* [40] *and the calculations by Shimizu* [24].

The energy and angular distributions of gamma rays transmitted through aluminum slabs were measured by Dardis and Scofield [41].

Figure 5.27 shows the results of comparing theoretical energy angular distributions and experimental ones obtained by Dardis and Scofield.

Agreement between measurement and calculation is not as good as that between the Monte Carlo calculation by Berger and Morris [36] and the invariant embedding calculations as shown in Fig. 5.22, but the shape of the spectrum agrees fairly well.

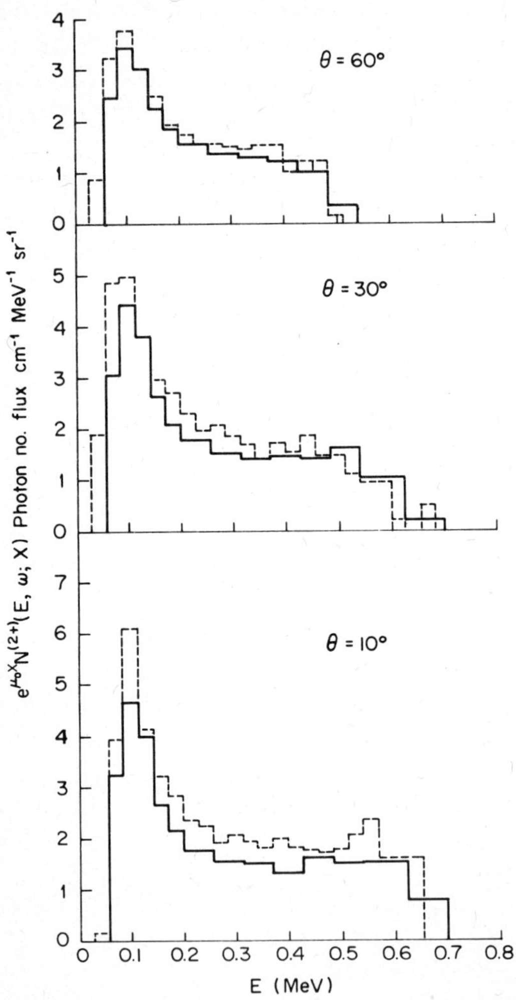

Fig. 5.27 *Energy spectra of radiations of the second- and higher order of scattering from a* ^{137}*Cs plane perpendicular source transmitted through an aluminum slab of* 3.080 *mean free paths at angles of* 10, 30, *and* 60°. – – – *experiments by Dardis and Scofield* [41], — *invariant embedding calculations* [24].

5.3 Transmission through Multilayer Slabs

5.3.1 Equation of Synthesis

The reflection and transmission functions of any multilayer slab can be synthesized from the reflection and transmission functions of each elementary layer. Suppose we have a two-layer slab consisting of a layer of material 1 and a layer of material 2. Let the reflection and transmission functions of the first and second layers be $\mathbf{R}_1$, $\mathbf{T}_1$ and $\mathbf{R}_2$, $\mathbf{T}_2$, respectively. We assume that the radiations of the unit intensity are incident on the first layer. Then the resultant reflected and transmitted currents from the two-layer slab are given in terms of the reflection and transmission functions of the two-layer slab, which we shall denote by $\mathbf{R}_{1+2}$ and $\mathbf{T}_{1+2}$, respectively. They are expressed as functions of the elementary layers by the equations

$$\mathbf{R}_{1+2} = \mathbf{R}_1 + \sum_{n=0}^{\infty} \mathbf{T}_1(\mathbf{R}_2 \cdot \mathbf{R}_1)^n \mathbf{R}_2 \mathbf{T}_1, \tag{5.15}$$

$$\mathbf{T}_{1+2} = \sum_{n=0}^{\infty} \mathbf{T}_2(\mathbf{R}_1 \mathbf{R}_2)^n \mathbf{T}_1. \tag{5.16}$$

The term $\mathbf{T}_2(\mathbf{R}_1\mathbf{R}_2)^n\mathbf{T}_1$ in Eq. (5.16) represents the transmitted radiations which have crossed the interface between the first and second layers $(2n + 1)$ times $[(n + 1)$ times in the forward direction and n times in the backward direction$]$. Similarly, the term $\mathbf{T}_1(\mathbf{R}_2\mathbf{R}_1)^n\mathbf{R}_2\mathbf{T}_1$ in Eq. (5.15) represents the reflected radiations which have crossed the interface $2(n + 1)$ times $[(n + 1)$ times in the forward direction and $(n + 1)$ times in the backward direction$]$. Equations (5.15) and (5.16) can be reduced to

$$\mathbf{R}_{1+2} = \mathbf{R}_1 + \mathbf{T}_1[\mathbf{E} - \mathbf{R}_2\mathbf{R}_1]^{-1}\mathbf{R}_2\mathbf{T}_1, \tag{5.17}$$

$$\mathbf{T}_{1+2} = \mathbf{T}_2[\mathbf{E} - \mathbf{R}_1\mathbf{R}_2]^{-1}\mathbf{T}_1, \tag{5.18}$$

where $\mathbf{A}^{-1}$ represents the inverse of the matrix $\mathbf{A}$, and $\mathbf{E}$ the unit matrix. Equations (5.17) and (5.18) are the basic equations for synthesizing the reflection and transmission functions of a composite slab. By repeated applications of these equations, the reflection and transmission functions

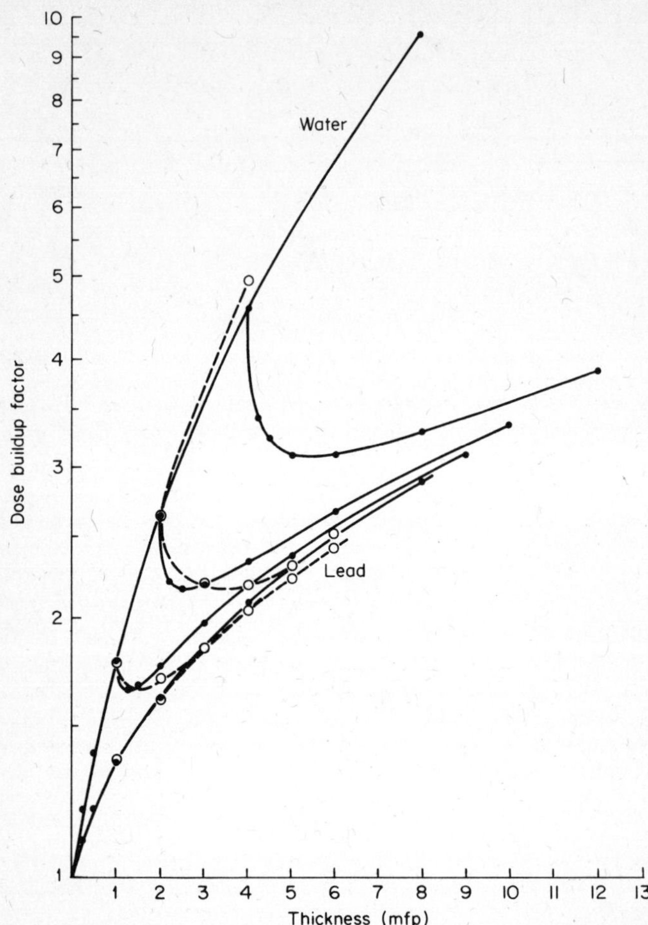

Fig. 5.28 *Dose buildup factor of water followed by lead for a plane perpendicular* 1.0-*MeV source.* ○ *Monte Carlo calculations by Bowman and Trubey* [42], ● *invariant embedding calculations* [25].

of any multilayer slab can be synthesized from the functions of each elementary layer. For example, the functions of a three-layer slab $\mathbf{R}_{1+2+3}$, $\mathbf{T}_{1+2+3}$ can be synthesized according to Eqs. (5.17) and (5.18) from the functions of the first layer $\mathbf{R}_1$, $\mathbf{T}_1$ and the functions of a composite slab consisting of the second and third layers $\mathbf{R}_{2+3}$, $\mathbf{T}_{2+3}$ which in turn can be synthesized from the functions of the second layer $\mathbf{R}_2$, $\mathbf{T}_2$ and those of the third layer $\mathbf{R}_3$, $\mathbf{T}_3$.

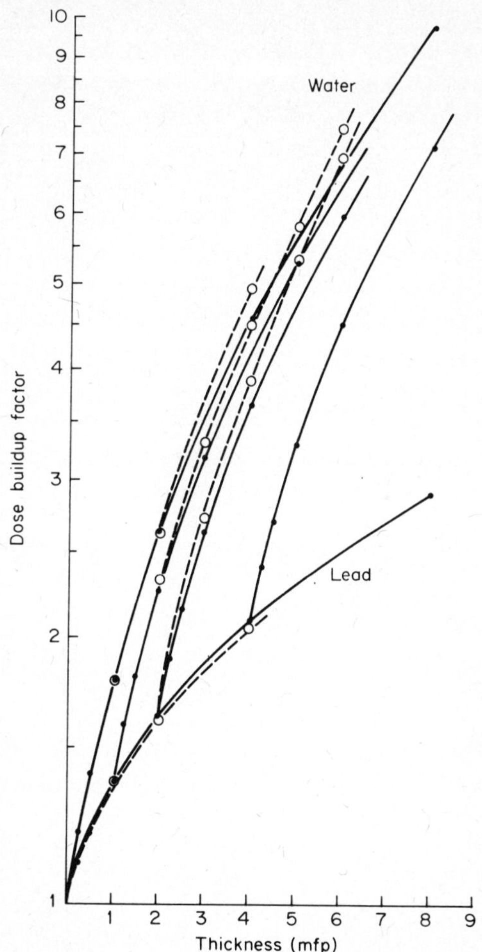

Fig. 5.29 *Dose buildup factor of lead followed by water for a plane perpendicular* 1.0-*MeV source.* ○ *Monte Carlo calculations by Bowman and Trubey* [42], ● *invariant embedding calculations* [25].

An important merit of the present approach in which the solutions are given in terms of the reflection and transmission functions is that the solutions for single layers can be used directly to obtain the solutions for multilayer slabs. This is not the case in the Boltzmann approach in which the solutions are described by flux.

5.3.2 Invariant Embedding Calculations

A series of calculations was made on the penetration of gamma rays through two-layer slabs by Shimizu [25] using the reflection and transmission functions of single layers obtained by the method of invariant embedding. The dose transmission factors for two-layer slabs were tabulated [27]. Figures 5.28 and 5.29 show the dose buildup factors of two-layer slabs of water and lead calculated both by the method of invariant embedding and the Monte Carlo method. The calculations by the Monte Carlo method were performed by Bowman and Trubey [42]. In Fig. 5.28, the dose buildup factors for τ_1 mean free paths of water followed by τ_2 mean free paths of lead are plotted as a function of the total thickness $(\tau_1 + \tau_2)$. Figure 5.29 shows the dose buildup factors for lead followed by those for water. The dose buildup factors for homogeneous water and lead slabs are also plotted in the figures.

REFERENCES

1. H. A. BETHE and J. ASHKIN, "Experimental Nuclear Physics," Vol. 1. Wiley, New York, 1953.
2. W. HEITLAR, "Quantum Theory of Radiation." Oxford Univ. Press, New York and London, 1954.
3. U. FANO, L. V. SPENCER, and M. J. BERGER, *in* "Handbuch der Physik" (S. Flügge, ed.), Vol. 38/2. Springer, Berlin and New York, 1959.
4. H. GOLDSTEIN, "Fundamental Aspects of Reactor Shielding." Addison-Wesley, Reading, Massachusetts, 1959.
5. L. V. SPENCER and U. FANO, *Phys. Rev.* **81**, 464L (1951).
6. L. V. SPENCER and U. FANO, *J. Res. Nat. Bur. Std.* **46**, 446 (1951).
7. H. GOLDSTEIN and J. E. WILKINS, Jr., Calculations of penetration of gamma rays, NYO-3075 (1954).
8. M. J. BERGER and J. A. DOGGETT, Reflection and transmission of gamma radiation by barriers, *J. Res. Nat. Bur. Std.* **56**, 89 (1956).
9. C. H. PEEBLES and M. S. PLESSET, Transmission of gamma-rays through large thicknesses of heavy materials, *Phys. Rev.* **81**, 430 (1951).
10. R. ARONSON and D. L. YARMUSH, Transfer-matrix method for gamma-ray and neutron penetration, *J. Math. Phys.* (*N.Y.*) **7**, 221 (1966).
11. I. KATAOKA, Use of response matrix method for multi-layer shield design, *Proc. Int. Conf. Peaceful Uses At. Energy, 3rd* **4**, 386 (1965).
12. I. KATAOKA and K. TAKEUCHI, A method for the numerical integration of the photon transport equation in slab geometry, *J. Nucl. Sci. Technol.* **2**, 30 (1965).
13. K. D. LATHROP, Use of discrete-ordinates methods for solution of photon transport problems, *Nucl. Sci. Eng.* **24**, 381 (1966).
14. V. A. AMBARZUMIAN, *Compt. Rend. Acad. Sci. URSS* **38**, 299 (1943).
15. S. CHANDRASEKHAR, "Radiative Transfer." Oxford Univ. Press, London and New York, 1950.
16. R. BELLMAN and R. KALABA, On the principle of invariant embedding and propagation through inhomogeneous media, *Proc. Nat. Acad. Sci. USA* **42**, 629 (1956).
17. R. BELLMAN and R. KALABA, On the principle of invariant embedding and diffuse

reflection from cylindrical regions, *Proc. Nat. Acad. Sci. USA* **43**, 517 (1957). See also references 15–21 in Part B.

18. R. PREISENDORFER, A mathematical foundation for radiative transfer theory, *J. Math. Mech.* **6**, 685 (1957).

19. S. UENO, The Probabilistic method for problems of radiative transfer, X. Diffuse reflection and transmission in a finite inhomogeneous atmosphere, *Astrophys. J.* **132**, 729 (1960).

20. S. UENO, Stochastic equations in radiative transfer by invariant embedding method, *J. Math. Anal. and Appl.* **2**, 217 (1961).

21. R. BELLMAN, R. KALABA, and M. C. PRESTRUD, "Invariant Embedding and Radiative Transfer in Slabs of Finite Thickness." American Elsevier, New York, 1963.

22. A. SHIMIZU and H. MIZUTA, Application of invariant embedding to the reflection and transmission problem of gamma rays, I, *J. Nucl. Sci. Technol.* **3**, 57 (1966).

23. A. SHIMIZU and H. MIZUTA, Application of invariant embedding to the reflection and transmission problem of gamma rays, II, *J. Nucl. Sci. Technol.* **3**, 441 (1966).

24. A. SHIMIZU, Calculation of the penetration of gamma rays through slabs by the method of invariant embedding, *Nucl. Sci. Eng.* **32**, 184 (1968).

25. A. SHIMIZU, Calculation of the penetration of gamma rays through two-layer slabs, *Nucl. Sci. Eng.* **32**, 385 (1968).

26. A. SHIMIZU, Tabulation of dose transmission factors for homogeneous slabs, NBS Rep. 9617 (1967).

27. A. SHIMIZU, Tabulation of dose transmission factors for two-layer slabs, NBS Rep. 9618 (1967).

28. D. R. MATHEWS, K. F. HANSEN, and E. A. MASON, Deep penetration of radiation by the method of invariant embedding, *Nucl. Sci. Eng.* **27**, 263 (1967).

29. B. P. Bulatov and E. A. Garusov, ^{60}Co and ^{198}Au γ-ray albedo of various materials, *J. Nucl. Energy* **11**, 159 (1960).

30. H. FUJITA, K. KOBAYASHI, and T. HYODO, Backscattering of gamma rays from iron slabs, *Nucl. Sci. Eng.* **19**, 437 (1964).

31. M. J. BERGER and D. J. Raso, Backscattering of gamma rays, NBS Rep. 5982 (1960).

32. M. J. BERGER and D. J. RASO, *Radiation Res.* **12**, 20 (1960).

33. B. P. BULATOV, The albedos of various substances for γ rays from isotropic ^{60}Co, ^{137}Cs, and ^{51}Cr sources, *J. Nucl. Energy* **13**, 84 (1960).

34. T. HYODO, Backscattering of gamma rays, *Nucl. Sci. Eng.* **12**, 178 (1962).

35. M. J. BERGER and E. MORRIS, NBS 9071 (1966).

36. M. J. BERGER and E. MORRIS, Private communication (1966).

37. L. V. SPENCER and J. LAMKIN, Slant penetration of γ-rays in H_2O, NBS 5944 (1958).

38. M. J. BERGER, *J. Nucl. Med.* (to be published).

39. Y. FURUTA *et al.*, Dose buildup factors of plane parallel barriers for ^{60}Co plane monodirectional source, *Nucl. Sci. Eng.* **25**, 85 (1966).

40. K. TAMURA and A. TSURUO, Dose buildup factors of plane parallel barriers for ^{137}Cs plane monodirectional source, *Nucl. Sci. Eng.* **28**, 144 (1967).

41. J. G. DARDIS and N. E. SCOFIELD, Energy and angular distribution of gamma rays scattered in aluminum, *Phys. Rev.* **147**, 280 (1966).

42. L. A. BOWMAN and D. K. TRUBEY, Monte Carlo calculation of gamma-ray dose-rate buildup factors for lead and water shields, ORNL-2609, 110 (1958).

PART B *APPLICATION TO CRITICALITY CALCULATIONS*

CHAPTER SIX *INTRODUCTION*

6.1 Introduction to Criticality Calculations

Let us begin our study of "criticality calculations" by looking at the cross section of an actual reactor. Figure 6.1 shows the horizontal cross section of a "materials testing reactor" which was designed for the purpose of studying the physical properties of various materials under neutron irradiation. The reactor consists of fuel subassemblies, reflector subassemblies, control rods, and experimental holes. Each subassembly has its own fine structure. For example, a fuel subassembly consists of fuel plates and a water moderator.

Criticality calculations on such a reactor are usually performed in two steps: first the microscopic calculation and then the macroscopic one. The microscopic calculations consist of calculating the fine distribution of neutrons within a subassembly, taking the heterogeneous structures into account. The results of microscopic calculations are expressed in terms of the "group constants" for homogenized subassemblies. The macroscopic calculations consist of calculating the criticality and overall flux distribution in the reactor which is now considered as consisting of homogenized subassemblies.

In the rest of this monograph, we shall confine ourselves to the discussions of macroscopic calculations.

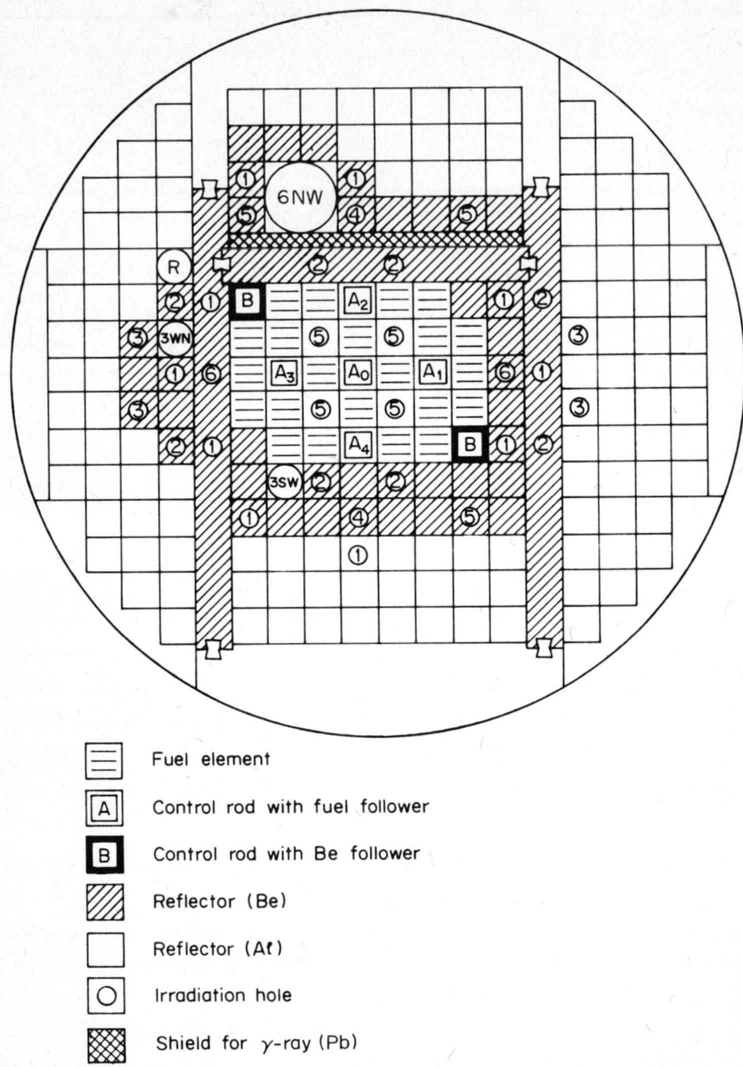

Fig. 6.1 *Horizontal cross section of a materials testing reactor.*

The design of a reactor usually requires a large amount of criticality calculations of various kinds. As an example, we shall show the types and numbers of criticality calculations required in the design of a materials testing reactor.*

* The authors owe the subsequent descriptions and tables in Section 6.1 to Mr. T. Omura of Nippon Atomic Industry Group Co., Ltd.

Table 6.1 *Specifications for a Materials Testing Reactor*

Condition	Item	Specifications
S1	Power	50 MW
S2	Number of experimental holes	$4 \sim 8$ fast neutron irradiation holes in the fuel region. As many as possible thermal neutron irradiation holes in the reflector region.
S3	Neutron flux	Fast neutron flux above 1 MeV $> 2 \times 10^{14}$ cm^{-2} sec^{-1}. Thermal neutron flux $> 2 \times 10^{14}$ cm^{-2} sec^{-1}.
S4	Days of continuous operation	>2 weeks (>3 weeks for 30-MW operation).

Table 6.1 gives the design specifications for the reactor. The objectives, types, and number of cases of criticality calculations performed for the conceptual design are given in Table 6.2, and those performed for the detailed design in Table 6.3.

From these tables, it can be seen that the two- or three-dimensional criticality calculations based on the few-group diffusion theory are performed very frequently and require much computer time. In concluding this section, we wish to emphasize that the development of efficient methods of criticality calculation (particularly, of two or three dimensions) is very important for making practical design calculations.

6.2 Classical Approach

In this section we shall briefly describe the conventional method of criticality calculations which consists of numerical solutions of group diffusion equations.

Table 6.2 *Criticality Calculations Required for the Conceptual Design*

	Objective	Variables	Types of criticality calculations[a]	Number of cases	Computation time on IBM-7090 (hours)[b]
C1	Determination of Core Volume To determine the core volume which satisfies conditions S1 and S2.	Number and position of irradiation holes. Number and length of fuel elements.	2-dim diffusion $G = 2$ or 3 [flux in the holes] [power peaking]	25	$25 \times 0.2 = 5$
C2	Determination of Fuel Loading For condition S4, relations between fuel loading and the required number of control rods are studied.	Fuel loading. Number and position of control rods.	2-dim diffusion $G = 2$ or 3 [maximum fuel loading satisfying $k_{eff} \leq 0.9$, when all control rods are inserted]	25	$25 \times 0.2 = 5$
			1-dim burnup [possible days of continuous operation] [effects of fuel shuffling]	50	$50 \times 0.05 = 2.5$
C3	Determination of Core Configuration To determine the core configuration, more refined calculations are performed based on the results of C1 and C2.		2-dim diffusion $G \geq 4$ with fine mesh	20	$20 \times 0.5 = 10$

C4	Void and Temperature Coefficients To check the safety of the reactor, the reactivity coefficients of voids and temperature are calculated. The results are used for the analysis of core kinetics.	2-dim diffusion 2-dim perturbation	20 20	$20 \times 0.2 = 4$ $20 \times 0.02 = 0.4$
C5	Effects of Control Rods Changes of fluxes in the irradiation holes, due to the movements of the control rods, are investigated. The effects of control rod movements on the power peaking factor should also be checked against the melting or burnout of fuel elements. Three-dimensional criticality calculations are necessary.	2-dim diffusion 3-dim diffusion (if possible) 3-dim diffusion (by flux synthesis)	30 5 10	$30 \times 0.2 = 6$? $10 \times 0.4 = 4$

[a] The outputs of the calculations are given in brackets.
[b] Standard computation time by the conventional codes.

Table 6.3 *Criticality Calculations Required for the Detailed Design*

Objectives	Types of criticality calculations	Number of cases	Computation time on IBM-7090 (hours)[a]
1. As the design proceeds, the fine structures of the reactor and all the component materials become concrete.	2-dim diffusion $G = 4$	30	$30 \times 0.5 = 15$
2. Data from the critical experiments are available at this stage.	3-dim diffusion (if possible)	10	?
Based on these conditions more refined calculations are performed for the final design.	3-dim diffusion (by flux synthesis)	10	$10 \times 0.4 = 4$
	2-dim burnup (~6 time steps)	10	$10 \times 2.5 = 25$

[a] Standard computation time by the conventional codes.

6.2.1 Group Diffusion Equation

The most widely used theory for macroscopic calculations is group diffusion theory, in which the neutron energy is represented by a finite number G of energy groups. Group diffusion theory is broadly classified into the "few-group diffusion theory" and the "multigroup diffusion theory." For criticality calculations of thermal reactors in which slow neutrons play the main part, the few-group theory (for example, $G < 6$) is well suited as a handy tool, whereas the multigroup theory may be required for more refined calculations or for the fast reactors in which fast neutrons play the main role.

The group diffusion equation can be obtained by considering the neutron conservation within each energy group in the unit volume element of space. The derivation of the equation will be found in standard textbooks on reactor physics [1, 2].

The equation, in the simplest form, can be written in the form

$$\begin{aligned}
-\operatorname{div}(D_1 \operatorname{grad} \phi_1(r)) + \Sigma_1\phi_1(r) &= \frac{1}{\lambda}\chi_1 S(r), \\
-\operatorname{div}(D_2 \operatorname{grad} \phi_2(r)) + \Sigma_2\phi_2(r) &= \Sigma_{s1}\phi_1(r) + \frac{1}{\lambda}\chi_2 S(r), \\
&\ \ \vdots \\
-\operatorname{div}(D_G \operatorname{grad} \phi_G(r)) + \Sigma_G\phi_G(r) &= \Sigma_{s\,G-1}\phi_{G-1}(r) + \frac{1}{\lambda}\chi_G S(r),
\end{aligned} \tag{6.1}$$

$$S(r) = \sum_{g=1}^{G} (\nu\Sigma_f)_g\phi_g(r). \tag{6.2}$$

The boundary conditions are

$\phi_g(r)$ *and* D_g grad $\phi_g(r)$ *are continuous at interfaces of spacial regions,* (6.3)

and

$\phi_g(r) + 2D_g(\text{grad}\ \phi_g(r))_n = 0$ *at the outer boundary of the reactor,* (6.4)

where the following definitions hold:

- r space variable,
- g index for energy groups,
- D_g diffusion coefficient,
- Σ_g removal cross section,
- Σ_{sg} slowing down cross section from group g to group $g+1$,
- $(\nu\Sigma_f)_g$ fission cross section times average number of neutrons liberated per fission,
- χ_g energy spectrum of fission neutrons ($\sum_{g=1}^{G} \chi_g = 1$),
- n represents the vector component outwardly normal to the boundary.

Here we have choosen the "static multiplication factor" λ (or equally k_{eff}) as the eigenvalue of the problem.

6.2.2 Finite Difference Method

The solution of the group diffusion equations may be obtained either by analytic methods or by direct numerical integrations. The analytic methods are applicable to some one-dimensional simple configurations. An example will be found in the textbook by Glasstone and Edlund [2]. Unfortunately, however, the analytic methods are hardly applicable to complex reactor configurations, particularly to two- or three-dimensional reactor configurations. The main difficulty stems from the fact that the coefficients of the equations vary from region to region.

The most widely used technique for the numerical solutions of diffusion equations is the finite difference method* which represents the whole reactor by a finite number of grid points and couples each grid point to its neighbors. Usually the grid points are arranged in a regular array as shown in Fig. 6.2. In this method the derivatives are approximated by finite differences between neighboring grid points.

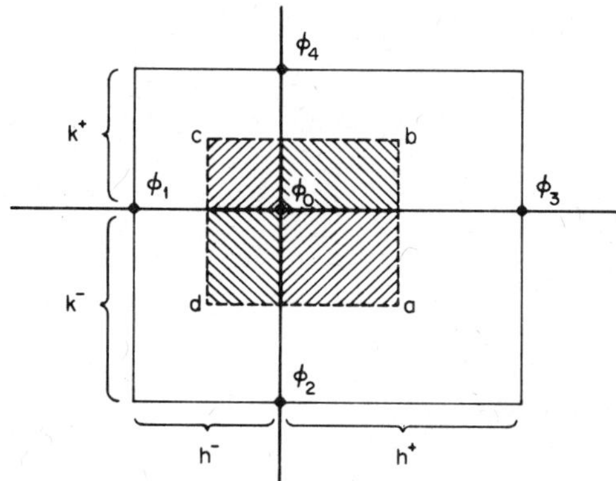

Fig. 6.2 *Derivation of five-point difference equation.*

In the following explanations, ϕ is assumed to be constant in each rectangle *abcd* centered at each grid point, while the coefficients (D, Σ, etc.) are assumed to be constant in each rectangle bounded by the neighboring four grid points.

* Detailed discussions of this method will be found in the textbook by Forsythe and Wasow [3]. Many applications to the reactor calculations are also described in recent books [4–6].

In view of the boundary condition (6.3), it is convenient to integrate Eq. (6.1) over the rectangle *abcd*, because by Gauss's theorem the first term reduces to the line integral along the boundary of the rectangle *abcd*:

$$\int \operatorname{div}\{D \operatorname{grad} \phi\}\, dv = \oint (D \operatorname{grad} \phi)_n\, dS, \tag{6.5}$$

where n represents the vector component outwardly normal to the boundary. The derivatives are then approximated by

$$\begin{aligned}
\frac{\partial \phi}{\partial x} &\doteqdot \frac{\phi_3 - \phi_0}{h^+} \qquad (\text{on } \overline{ab}), \\
\frac{\partial \phi}{\partial y} &\doteqdot \frac{\phi_4 - \phi_0}{k^+} \qquad (\text{on } \overline{bc}), \\
\frac{\partial \phi}{\partial x} &\doteqdot \frac{\phi_0 - \phi_1}{h^-} \qquad (\text{on } \overline{cd}), \\
\frac{\partial \phi}{\partial y} &\doteqdot \frac{\phi_0 - \phi_2}{k^-} \qquad (\text{on } \overline{da}).
\end{aligned} \tag{6.6}$$

The remaining terms can be integrated directly because we have assumed that ϕ is constant ($= \phi_0$) in *abcd*. They contain ϕ_0 as the only variable. Thus we obtain a five-point difference equation which relates ϕ_0 to ϕ_1, ϕ_2, ϕ_3, and ϕ_4.

Performing this procedure for all grid points (N points) and for all energy groups (G groups), we finally obtain simultaneous linear equations in $N \times G$ unknowns. These equations can be summarized by a matrix eigenvalue equation,

$$\mathbf{A}\boldsymbol{\phi} = \frac{1}{\lambda} \mathbf{M}\boldsymbol{\phi}, \tag{6.7}$$

where $\boldsymbol{\phi}$ is the vector representation of the $N \times G$ unknowns.

6.2.3 Outer and Inner Iterations

Equation (6.7) is usually solved by the iterative method. The basic scheme of the iteration procedure is

$$\mathbf{A}\boldsymbol{\phi}^{(t+1)} = \mathbf{M}\boldsymbol{\phi}^{(t)}, \qquad t = 0, 1, 2, \ldots, \tag{6.8}$$

where $\boldsymbol{\phi}^{(0)}$ is the initial-guess vector, and one may choose $\boldsymbol{\phi}^{(0)}$ as a vector with all its components equal to unity. It can be proved that $\boldsymbol{\phi}^{(t)}$ converges to the eigenfunction of (6.7) corresponding to the largest eigenvalue. The largest eigenvalue λ can be obtained from the relation

$$\boldsymbol{\phi}^{(t+1)} = \lambda \boldsymbol{\phi}^{(t)}. \tag{6.9}$$

This iterative process is usually called "outer iteration" (or source iteration).

Each outer iteration involves the process of matrix inversion; in other words, one must solve the matrix equation

$$\mathbf{AX} = \mathbf{K} \tag{6.10}$$

for a given vector **K**.

The Gauss elimination method is well suited to the one-dimensional problem, because in this case **A** becomes a block-tridiagonal matrix. But in the case of two- or three-dimensional problems, it is customary to resort to the iterative method. The iterative process for solving (6.10) is often called "inner iteration" (or flux iteration).

Many techniques for accelerating the convergence of inner and outer iterations have been extensively investigated, and some combinations of these have been realized in the design codes [7, 8].

Details of the iterative methods are beyond the scope of this monograph. They are summarized in recent books [4, 9].

6.2.4 Variational Flux Synthesis

Even in this age of modern high speed computers, it is still a formidable task to solve Eq. (6.7) and accordingly (6.10) for three-dimensional cases. Methods for synthesizing three-dimensional fluxes from one- and two-dimensional fluxes have been successfully developed [10, 11], based on the variational principle [12].

The variational synthesis is also applicable to burnup calculations or kinetics calculations for reactors. Application studies along these lines are being performed by various researchers.

For the purpose of providing a sketchy knowledge about the classical approach, we have very briefly described the conventional methods of numerical solution of group diffusion equations. Mention should be made here, however, that much effort is being exercised along these lines to improve the efficiency of the methods, and accordingly there are many books and papers available for reference.

6.3 Invariant Embedding Approach

The classical description of neutrons in a reactor is based on the choice of the flux and net current as the variables of interest. Alternative modes of description may, however, be possible.

In Chapters 7 and 8, we shall discuss some methods of criticality calculations which are based on the choice of the partial currents (the current parallel and antiparallel to the direction of net flow) or their reflection and transmission probabilities as the variables of interest.

In the field of neutron physics, "albedo" was first introduced by Amaldi and Fermi [13] for indicating the reflection power of a medium. The albedo introduced by them is identical to a one-velocity version of the reflection function defined in Chapter 2. It has been used in criticality calculations in a supporting role to simplify the calculation of core flux [2].

Recently, several authors proposed giving albedo a leading role in criticality calculations. Although the proposed methods were developed as extensions from different origins and written with one variation or another, they are essentially based on the same basic idea, which can be traced back to a paper on the reflection and transmission of light by piles of plates written almost one hundred years ago by Stokes [14].

Bellman *et al.* [15–21] proposed a method of criticality calculations based on the invariant embedding principle introduced in the field of astrophysics. They express the criticality condition in terms of the reflection function. According to their method, the critical thickness of a reactor can be obtained by solving the reflection function equation for increasing thickness. Their method will be described in Section 7.2.

Their method is also applicable to time-dependent transport problems. A series of calculations was made for time-dependent transport problems by Bellman *et al.* [22].

Ackroyd and McCullen [23] proposed a method based on the criticality condition that the core albedo should be equal to the effective reflector albedo.

They made actual calculations with the multigroup approximation, yet their criticality condition is essentially based on the one-group approximation. Otsuka [24] derived the criticality condition in terms of the core and reflector albedos with the two-group approximation. In their formulations, Ackroyd *et al.* and Otsuka used only the albedos. Accordingly, their methods are applicable to simple reactors consisting of a core and a reflector but not to multiregion reactors.

Selengut [25] developed a partial current representation as an extension of some early work by Bobrowsky [26] in connection with radiation shielding. Selengut describes regions in a reactor in terms of the reflection and transmission coefficients of partial current and derives the criticality condition in terms of those coefficients with the one-group approximation.

One of the authors [27] also developed a method of criticality calculations. His method is essentially the same as that proposed by Selengut except that the former is formulated using the multigroup approximation.

Actual calculations were made on slab reactors by Monta, Miyamoto, and one of the authors [28].

The method was further extended to two-dimensional problems by the authors [29]. This method and its applications will be given in Chapters 7 and 8.

It should be mentioned here that a matrix formulation of the group diffusion theory proposed by Schmid [30] and a generalized potential theory developed by Koskinen [31] have strong connections with the above-mentioned methods, although their formulations do not explicitly use the reflection and transmission functions. The transfer matrix introduced by Schmid is equal to the reflection and transmission matrices to be used here, provided a suitable transformation is made.

Auerbach [32] applied the method of invariant embedding developed by Chandrasekhar [33] to several problems in reactor theory.

The reflection and transmission of neutrons were also discussed by Wachspress [34] and Wilf [35].

CHAPTER SEVEN *ONE-DIMENSIONAL PROBLEM*

7.1 Direct Application of Invariant Embedding to Criticality Calculations

In this section, the criticality condition of a slab reactor will be derived in terms of the reflection function. By making use of this criticality condition, the critical size of the reactor can be obtained by calculating only the reflection function; calculation of the internal flux distribution is not necessary.

7.1.1 Criticality Condition in Terms of the Reflection Function

A critical reactor is one that can sustain the neutron density constant in time without any supply of neutrons from external sources. This means that if neutrons were injected constantly from external sources into a critical reactor, the resulting neutron density in the reactor would increase without limit with time. Therefore, the expected number of neutrons reflected by or transmitted through a critical reactor due to incident neutrons constant in time becomes infinite, that is, the reflection or transmission function of a critical reactor viewed from the outer surface is infinite.

Consider a slab reactor of thickness X. Let the reflection function of the reactor viewed from the outer surface be $\mathbf{R}(X)$. Then the critical thickness of the reactor X_c can be determined from the condition

$$\begin{aligned} &\mathbf{R}(X) \quad \text{is finite for} \quad X < X_c, \\ &\mathbf{R}(X) \quad \text{is infinite at} \quad X = X_c. \end{aligned} \tag{7.1}$$

The criticality calculation is thus reduced to a calculation of the reflection function for increasing thickness. This procedure for the criticality calculation has been developed by Bellman, Kalaba, and Wing. Their method permits us to determine the criticality without making any calculation of the internal flux distribution within the reactor. In other words, one may regard the reactor as "a black box," a box which is not open to us. In order to obtain the critical size, one need not know what happens inside the box. It is sufficient to observe the outputs from the box (the reflected or transmitted neutrons) due to a given input (the incident neutrons). The relation between the input and the outputs is expressed in terms of the reflection and the transmission functions.

As will be seen later, the internal flux distribution, which is often desired for reactor design, can also be obtained from the reflection and the transmission functions. A difference between the classical approach and the invariant embedding approach described here is that for the latter we can obtain the criticality first and then the internal flux if desired, whereas for the former we cannot obtain the criticality until we calculate the internal flux.

7.1.2 Equations for Reflection and Transmission Functions for Neutrons

The equations for the reflection and transmission functions derived for photons in Section 2.1 are valid for neutrons too. Some modification, however, is necessary, when the medium contains fissionable elements.

In this case, the functions $\dot{R}(E, \omega \mid E_0, \omega_0)$ and $\dot{T}(E, \omega \mid E_0, \omega_0)$ given by Eqs. (2.11) and (2.12) should be modified to the forms

$$\dot{R}(E, \omega \mid E_0, \omega_0) = \frac{1}{\omega_0} \Sigma_s(E_0, \omega_0 \to E, -\omega) + \frac{1}{2\omega_0} \chi(E)\nu\Sigma_f(E_0), \tag{7.2}$$

$$\dot{T}(E, \omega \mid E_0, \omega_0) = \delta(E - E_0)\, \delta(\omega - \omega_0) \frac{\Sigma(E)}{\omega} - \frac{1}{\omega_0} \Sigma_s(E_0, \omega_0 \to E, \omega) - \frac{1}{2\omega_0} \chi(E) \nu \Sigma_f(E_0), \tag{7.3}$$

where $\chi(E)$ represents the normalized fission spectrum, ν the average number of neutrons liberated per fission, and $\Sigma_f(E)$ the macroscopic fission cross section. It is assumed in the derivation of Eqs. (7.2) and (7.3) that neutrons are liberated isotropically by fission.

By inserting the expressions (7.2) and (7.3) into Eqs. (2.19)–(2.22) of Section 2.1, one obtains the desired equations for a homogeneous slab. They are

$$\begin{aligned} \frac{\partial}{\partial X} R(E, \omega \mid E_0, \omega_0; X) &= \int_0^\infty dE' \int_0^1 d\omega'\, T(E, \omega \mid E', \omega'; X) \\ &\times \int_0^\infty dE'' \int_0^1 \frac{d\omega''}{\omega''} [\Sigma_s(E'', \omega'' \to E', -\omega') + \tfrac{1}{2}\chi(E')\nu\Sigma_f(E'')] \\ &\times T(E'', \omega'' \mid E_0, \omega_0; X), \end{aligned} \tag{7.4}$$

$$\begin{aligned} \frac{\partial}{\partial X} T(E, \omega \mid E_0, \omega_0; X) &= -\frac{\Sigma(E)}{\omega} T(E, \omega \mid E_0, \omega_0; X) \\ &+ \int_0^\infty dE' \int_0^1 \frac{d\omega'}{\omega'} [\Sigma_s(E', \omega' \to E, \omega) \\ &+ \tfrac{1}{2}\chi(E)\nu\Sigma_f(E')] T(E', \omega' \mid E_0, \omega_0; X) \\ &+ \int_0^\infty dE' \int_0^1 d\omega'\, R(E, \omega \mid E', \omega'; X) \\ &\times \int_0^\infty dE'' \int_0^1 \frac{d\omega''}{\omega''} [\Sigma_s(E'', \omega'' \to E', -\omega') \\ &+ \tfrac{1}{2}\chi(E')\nu\Sigma_f(E'')] T(E'', \omega'' \mid E_0, \omega_0; X), \end{aligned} \tag{7.5}$$

$$
\begin{aligned}
&\frac{\partial}{\partial X} R(E, \omega \mid E_0, \omega_0; X) \\
&\quad = \frac{1}{\omega_0} \Sigma_s(E_0, \omega_0 \to E, -\omega) - \left[\frac{\Sigma(E)}{\omega} + \frac{\Sigma(E_0)}{\omega_0}\right] R(E, \omega \mid E_0, \omega_0; X) \\
&\quad + \int_0^\infty dE' \int_0^1 \frac{d\omega'}{\omega'} [\Sigma_s(E', \omega' \to E, \omega) \\
&\quad + \tfrac{1}{2}\chi(E)\nu\Sigma_f(E')] R(E', \omega' \mid E_0, \omega_0; X) \\
&\quad + \frac{1}{\omega_0} \int_0^\infty dE' \int_0^1 d\omega' \, R(E, \omega \mid E', \omega'; X) \\
&\quad \times [\Sigma_s(E_0, \omega_0 \to E', \omega') + \tfrac{1}{2}\chi(E')\nu\Sigma_f(E_0)] \\
&\quad + \int_0^\infty dE' \int_0^1 d\omega' \, R(E, \omega \mid E', \omega'; X) \\
&\quad \times \int_0^\infty dE'' \int_0^1 \frac{d\omega''}{\omega''} [\Sigma_s(E'', \omega'' \to E', -\omega') + \tfrac{1}{2}\chi(E')\nu\Sigma_f(E'')] \\
&\quad \times R(E'', \omega'' \mid E_0, \omega_0; X),
\end{aligned}
\tag{7.6}
$$

$$
\begin{aligned}
&\frac{\partial}{\partial X} T(E, \omega \mid E_0, \omega_0; X) \\
&\quad = -\frac{\Sigma(E_0)}{\omega_0} T(E, \omega \mid E_0, \omega_0; X) \\
&\quad + \frac{1}{\omega_0} \int_0^\infty dE' \int_0^1 d\omega' \, T(E, \omega \mid E', \omega'; X) \\
&\quad \times [\Sigma_s(E_0, \omega_0 \to E', \omega') + \tfrac{1}{2}\chi(E')\nu\Sigma_f(E_0)] \\
&\quad + \int_0^\infty dE' \int_0^1 d\omega' \, T(E, \omega \mid E', \omega'; X) \\
&\quad \times \int_0^\infty dE'' \int_0^1 \frac{d\omega''}{\omega''} [\Sigma_s(E'', \omega'' \to E', -\omega') + \tfrac{1}{2}\chi(E'')\nu\Sigma_f(E')] \\
&\quad \times R(E'', \omega'' \mid E_0, \omega_0; X).
\end{aligned}
\tag{7.7}
$$

7.1.3 Differences between Penetration Calculation and Criticality Calculation

The fission term $(1/2\omega_0)\chi(E)\nu\Sigma_f(E_0)$ in Eq. (7.2) modifies the form of Eqs. (7.4)–(7.7) only slightly, yet it gives rise to a number of difficulties in obtaining solutions. In fact, any solution of Eqs. (7.4)–(7.7) based on some realistic model has not yet been obtained.

The fission has a pronounced effect on the formation of the energy spectrum of neutrons. Through the fission process, neutrons, which are otherwise invariably slowed down, may reemerge with energy higher than that before collision. Consequently, we cannot obtain the flux at any energy until we calculate fluxes at all energies simultaneously. This should be compared with the penetration calculations, in which the energy spectrum of neutrons or photons can be calculated successively from the higher to the lower energy.

Furthermore, when the medium contains fissionable elements of more than a certain concentration, the asymptotic value of the reflection function becomes infinite. Consequently, the method of solutions based on the use of the modified transmission function which is quite useful for the penetration calculations is not applicable to criticality calculations. Thus, the fission process makes it difficult to solve Eqs. (7.4)–(7.7).

On the other hand, since neutrons in a reactor core are almost isotropic, the treatment of the angular distribution is much simpler than that required for highly anisotropic radiations penetrating a shield. In fact, most criticality calculations have been made based on the diffusion approximation, as far as the overall flux distribution and the associated effective multiplication factor are concerned.*

7.2 Modification of Invariant Embedding Method

As mentioned in the previous section, the diffusion approximation is applicable to most criticality calculations. On the basis of diffusion theory, the criticality factor and the neutron distribution in a homogeneous bare reactor are readily obtainable by using the elegant theory developed by Weinberg and Wigner [1].

It is the nonuniformity of real reactors that makes diffusion calculations difficult. Furthermore, real reactors are of two- or three-dimensional geometry as well as nonuniform. Any method of criticality calculations which is applicable only to one-dimensional reactors is not very useful.

In Sections 7.2-7.5 and Chapter 8, we shall discuss primarily a modified application of invariant embedding to criticality calculations, instead

* It is necessary, however, to use more refined theories, if one is concerned with the fine distribution in a heterogeneous cell.

of the direct method based on the direct integration of Eq. (6.1). It should, however, be mentioned that it is still worthwhile obtaining any solution of Eq. (6.1) based on some realistic model.

The modified method of criticality calculations developed by the authors [27–29] consists of a combination of the "black box" concept, underlying the invariant embedding approach, and diffusion theory. The method will be called the "response matrix method."

7.2.1 Response Matrix Method

The response matrix method consists of the response matrices for each spatial region, the law of synthesis for these matrices, and the criticality equation in terms of these matrices.

The outline of the method is shown schematically in Fig. 7.1 as compared to the classical method based on the direct numerical integration of the diffusion equation for a reactor.

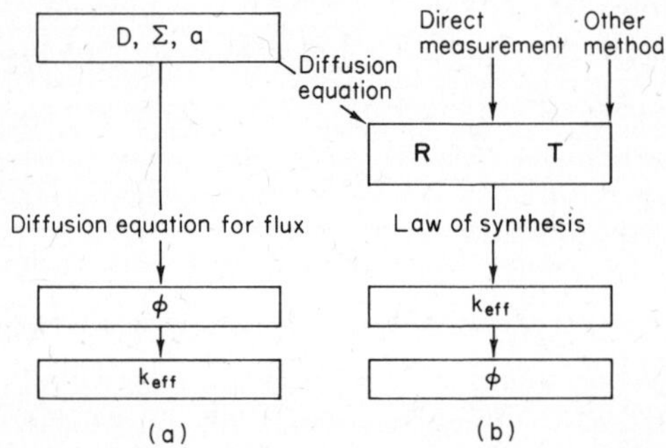

Fig. 7.1 *Outline of the response matrix method. (a) Classical method; (b) response matrix method.*

In the response matrix method the reactor is assumed to be divided into a number of homogeneous regions. Each region, in turn, is regarded as a black box. Its nuclear characteristics are represented by the response matrices which are identical to a multigroup version of the reflection and transmission functions. The response matrices may be obtained either by diffusion theory, or by more refined theories or by direct measurements, if necessary. Once the response matrices are obtained, the criticality can be determined by solving the criticality equation with the aid of the law of synthesis, which involves only some operations on those matrices. Internal fluxes are then obtainable if required.

The procedure adopted here has the feature that the criticality calculation of a multiregion reactor is divided into two steps: the calculation of response matrices for component regions and the criticality calculation in terms of these matrices. It has the following advantages over direct integration of the diffusion equation for a multiregion reactor:

(1) When the few-group diffusion theory is employed, the present procedure is much more efficient than direct numerical integration of the diffusion equation, since the response matrices can be calculated analytically and the criticality calculation involves only some operations on those matrices of low order. The reduction in computation time is more than a factor of 10 as compared to the direct numerical integration.

(2) Treatment of a region such as a control plate or a cavity, to which a successful application of diffusion approximation may be difficult, becomes easier because once the response matrices of such regions are obtained by a suitable theory, they can be used together with the matrices of other regions for criticality calculation.

(3) The response matrices have a clear physical meaning and, in principle, it can be measured directly.

7.2.2 Basic Assumptions

According to the present method, any spatial region is regarded as a black box which responds to the incident current of neutrons and gives rise to reflected and transmitted currents. The response is now assumed to satisfy the following two conditions:

(i) the response is linear,
(ii) the response is stationary.

Condition (i) means that "the principle of superposition" is applicable to the response. Condition (ii) means that a response current at time t to a pulsed incident current at time t_0 depends only on the time interval $(t - t_0)$.

Both conditions are satisfied as far as the properties of the medium, including temperature, are unchanged in time and by neutron irradiations. In performing the static criticality calculation of any reactor and the dynamic (in other words, kinetic) calculations of zero-power reactors, the properties of the medium may be considered as unchanged. However, when the dynamic performances of a power reactor are concerned, the variation of the temperature of the medium through fission reactions plays an essential role. Such a problem requires a suitable modification of the present method and is beyond the scope of the present discussion.

7.2.3 Response Matrices

The response matrices to be used here are versions of the reflection and the transmission functions based on the multigroup diffusion approximation.

According to the multigroup approximation, the energy range of interest is divided into a number of groups. For each energy group, we assign a standard energy distribution of incident current. Let the standard distribution of group n be $S_n(E, \omega)$. It is assumed to satisfy the normalization condition

$$\int_n dE \int_0^1 d\omega \, S_n(E, \omega) = 1. \tag{7.8}$$

Suppose the incident current of group m with total intensity J_m^+ having the standard distribution $S_m(E, \omega)$ is incident on a slab continuously in time. Then the total intensity of the reflected current of group n due to this incident current may be given by $R_{nm}J_m^+$ and that of the transmitted current by $T_{nm}J_m^+$. The proportionality constants R_{nm} and T_{nm} are expressed in terms of the reflection and transmission functions of the slab as

$$R_{nm} = \int_n dE \int_0^1 d\omega \int_m dE_0 \int_0^1 d\omega_0 \, R(E, \omega \mid E_0, \omega_0; X) S_m(E_0, \omega_0), \tag{7.9}$$

$$T_{nm} = \int_n dE \int_0^1 d\omega \int_m dE_0 \int_0^1 d\omega_0 \, T(E, \omega \mid E_0, \omega_0; X) S_m(E_0, \omega_0). \tag{7.10}$$

The square matrices **R** and **T** whose (n, m) components are R_{nm} and T_{nm}, respectively, will be called the "response matrices" of the slab. The order of the response matrices is equal to the number of energy groups.

The definition of the response matrices stated above is somewhat ambiguous, since the form of the standard distribution $\{S_n(E, \omega)\}$ is not given explicitly. The explicit specification of the standard distribution is necessary in order to measure the response matrices or to calculate them by transport theory. It is not necessary, however, for the calculations of the response matrices with the multigroup diffusion approximation.

In making such calculations, it is sufficient to assume that neutron flux obeys the multigroup diffusion equation with suitable group constants. Then the partial currents are expressed in terms of flux and its derivative as

$$J_n^{\pm}(X) = \frac{1}{4}\left\{\phi_n(X) \mp 2D_n \frac{d}{dX}\, \phi_n(X)\right\}, \tag{7.11}$$

where $J_n^+(X)$, $J_n^-(X)$, $\phi_n(X)$, and D_n are, respectively, the forward current, backward current, neutron flux, and the diffusion coefficient of the nth energy group.

The present formulation presumes that the forward and the backward currents are continuous at the interface between two regions. These conditions are equivalent to the conditions usually used in the diffusion calculations that flux and net current are continuous [cf. Eqs. (6.3) and (6.4)].

For inhomogeneous slabs or for regions of other one-dimensional geometries, a distinction should be made between the response matrices viewed from the left surface (or the inner surface in the case of an annulus) which will be denoted by $\mathbf{R}^+$ and $\mathbf{T}^+$, and those viewed from the right surface (or the outer surface), $\mathbf{R}^-$ and $\mathbf{T}^-$.

7.2.4 Law of Synthesis

Let us consider a composite slab consisting of two slabs A and B as shown in Fig. 7.2. Suppose an incident current $\mathbf{J}^+(0)$ is incident on the composite slab through the left free surface, where $\mathbf{J}^+(0)$ is a column vector, the mth component of which represents the intensity of the incident current of group m. Denote the induced currents at the interface

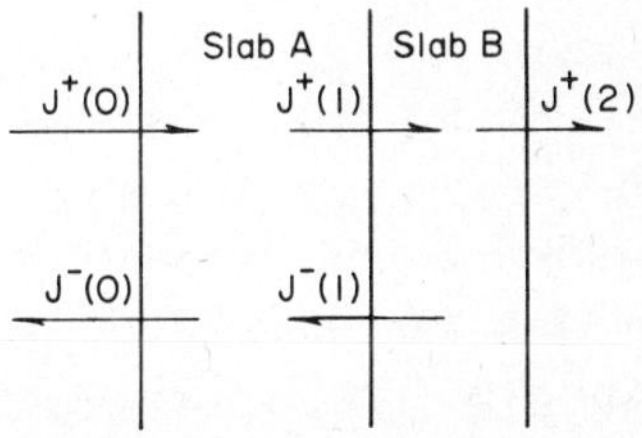

Fig. 7.2 *Law of synthesis.*

between A and B by $\mathbf{J}^{\pm}(1)$, the reflected current by $\mathbf{J}^{-}(0)$, and the transmitted current by $\mathbf{J}^{+}(2)$ (cf. Fig. 7.2).

Then, for slab A, one obtains the expressions

$$\mathbf{J}^{-}(0) = \mathbf{R}_A{}^{+}\mathbf{J}^{+}(0) + \mathbf{T}_A{}^{-}\mathbf{J}^{-}(1), \tag{7.12}$$

$$\mathbf{J}^{+}(1) = \mathbf{T}_A{}^{+}\mathbf{J}^{+}(0) + \mathbf{R}_A{}^{-}\mathbf{J}^{-}(1), \tag{7.13}$$

where $\mathbf{R}_A{}^{\pm}$ and $\mathbf{T}_A{}^{\pm}$ are the response matrices of slab A.

For slab B, we have

$$\mathbf{J}^{-}(1) = \mathbf{R}_B{}^{+}\mathbf{J}^{+}(1), \tag{7.14}$$

$$\mathbf{J}^{+}(2) = \mathbf{T}_B{}^{+}\mathbf{J}^{+}(1). \tag{7.15}$$

In the derivations of Eqs. (7.12)–(7.15) it is assumed that the distribution of the current within a group is kept unchanged from the standard distribution, even after transmitted or reflected by slabs. This assumption can be made either if the actual distribution within a group does not deviate much from the standard one, or if the response matrices are rather insensitive to the variation of the distribution within a group.

By eliminating $\mathbf{J}^{\pm}(1)$ from Eqs. (7.12)–(7.15), we obtain

$$\mathbf{J}^{-}(0) = \mathbf{R}_{A+B}^{+}\mathbf{J}^{+}(0), \tag{7.16}$$

$$\mathbf{J}^{+}(2) = \mathbf{T}_{A+B}^{+}\mathbf{J}^{+}(0), \tag{7.17}$$

where

$$\mathbf{R}_{A+B}^{+} = \mathbf{R}_A{}^{+} + \mathbf{T}_A{}^{-}(\mathbf{E} - \mathbf{R}_B{}^{+}\mathbf{R}_A{}^{-})^{-1}\mathbf{R}_B{}^{+}\mathbf{T}_A{}^{+}, \tag{7.18}$$

$$\mathbf{T}_{A+B}^{+} = \mathbf{T}_B{}^{+}(\mathbf{E} - \mathbf{R}_A{}^{-}\mathbf{R}_B{}^{+})^{-1}\mathbf{T}_A{}^{+}, \tag{7.19}$$

where $\mathbf{E}$ is the unit matrix and $\mathbf{A}^{-1}$ is the inverse matrix of $\mathbf{A}$. The matrices $\mathbf{R}_{A+B}^{+}$ and $\mathbf{T}_{A+B}^{+}$ are the response matrices of the composite slab viewed from the left surface.

Similarly, those matrices viewed from the right surface are expressed as

$$\mathbf{R}_{A+B}^{-} = \mathbf{R}_B{}^{-} + \mathbf{T}_B{}^{+}(\mathbf{E} - \mathbf{R}_A{}^{-}\mathbf{R}_B{}^{+})^{-1}\mathbf{R}_A{}^{-}\mathbf{T}_B{}^{-}, \tag{7.20}$$

$$\mathbf{T}_{A+B}^{-} = \mathbf{T}_A{}^{-}(\mathbf{E} - \mathbf{R}_B{}^{+}\mathbf{R}_A{}^{-})^{-1}\mathbf{T}_B{}^{-}. \tag{7.21}$$

Equations (7.18)–(7.21) represent the law of synthesis for the response matrices.

The response matrices of a composite slab consisting of more than two slabs can be obtained by the successive use of the law of synthesis.

7.2.5 Criticality Equation

From the viewpoint of a "black box," a reactor consisting of many regions corresponds to a combined system of black boxes.

A critical reactor is equivalent to a combined system capable of maintaining the stationary current by itself. The critical condition may be expressed in the following way.

For convenience, we divide a slab reactor into two subassemblies of spatial regions by a suitable interface.

Suppose a steady incident current of neutrons $\mathbf{J}^+$ is introduced into the right subassembly through the dividing interface.

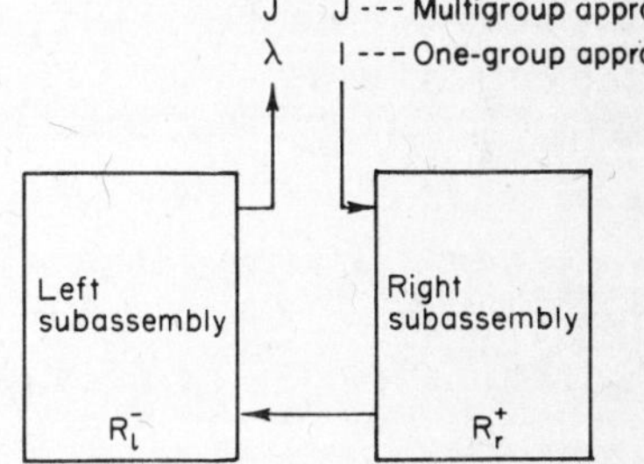

Fig. 7.3 *Schematic for the derivation of the criticality equation.*

It is assumed, as shown in Fig. 7.3, that the neutrons introduced will reemerge from the left subassembly with current density $\tilde{\mathbf{J}}^+$ after being reflected once by the right subassembly and then once by the left one. An appropriate selection of the spectrum (the relative intensity among the groups) of the incident current $\mathbf{J}^+$ will make the following relation hold:

$$\tilde{\mathbf{J}}^+ = \lambda \mathbf{J}^+, \tag{7.22}$$

where λ is a scalar constant.

From a "neutron life cycle" point of view that neutrons begin their new generation by entering the right subassembly and end it by leaving the left one as shown in Fig. 7.3, Eq. (7.22) may be interpreted as representing the spectrum of neutrons after passing through a generation as being the same as that at the beginning, and the population is λ times as large as the original.

Thus, the constant λ is the multiplication factor per generation, and the current satisfying Eq. (7.22) is the characteristic spectrum of the reactor.

Let the response matrices of the right and the left subassemblies viewed from the dividing interface be $\mathbf{R}_r^+$ and $\mathbf{R}_l^-$, respectively. Then, we obtain

$$\tilde{\mathbf{J}}^+ = \mathbf{R}_l^- \mathbf{R}_r^+ \mathbf{J}^+. \tag{7.23}$$

From Eqs. (7.22) and (7.23), the following eigenvalue equation is derived:

$$(\mathbf{R}_l^- \mathbf{R}_r^+ - \lambda \mathbf{E})\mathbf{J}^+ = 0. \tag{7.24}$$

The eigenvalue λ is determined by solving the secular equation

$$| \mathbf{R}_l^- \mathbf{R}_r^+ - \lambda \mathbf{E} | = 0. \tag{7.25}$$

The criticality condition is expressed as

$$\lambda \quad \text{(the maximum eigenvalue)} = 1. \tag{7.26}$$

To illustrate the physical meaning of the multiplication factor λ, we shall discuss a subcritical reactor with a plane source at the location of the dividing interface. The plane source is assumed to emit neutrons constantly into the right direction with intensity $\mathbf{J}_0^+$. Then the resulting stationary currents at the dividing interface $\mathbf{J}^\pm$ are expressed as

$$\mathbf{J}^+ = (\mathbf{E} - \mathbf{R}_l^- \mathbf{R}_r^+)^{-1} \mathbf{J}_0^+, \tag{7.27}$$

$$\mathbf{J}^- = \mathbf{R}_r^+ (\mathbf{E} - \mathbf{R}_l^- \mathbf{R}_r^+)^{-1} \mathbf{J}_0^+. \tag{7.28}$$

If the spectrum of $\mathbf{J}_0^+$ is identical to that of the eigenvector of Eq. (7.24), Eqs. (7.27) and (7.28) are reduced to

$$\mathbf{J}^+ = \frac{1}{1 - \lambda} \mathbf{J}_0^+, \tag{7.29}$$

$$\mathbf{J}^- = \frac{1}{1 - \lambda} \mathbf{R}_r^+ \mathbf{J}_0^+. \tag{7.30}$$

Equation (7.29) indicates that the reactor amplifies the source neutrons by the factor $1/(1 - \lambda)$.

It should be mentioned here that the multiplication factor λ introduced here is not identical to the multiplication factor introduced in Section 6.2 [cf. Eq. (6.1)]. The difference between both factors originates from the difference in the definition of the neutron life cycle. It should be noted further that both multiplication factors are called "static criticality factors" [1]. Strictly speaking, these factors and the corresponding characteristic distributions have their physical meanings only when the reactor is critical.

The rigorous treatment of time-dependent problems for noncritical reactors by the present formulation will be given in Appendix A.

7.2.6 Internal Flux

Once the current at the dividing interface is obtained by solving Eq. (7.24), the forward and the backward currents at the interfaces between spatial regions can be calculated by using the response matrices. This means that flux and its derivative at the interfaces can be obtained [cf. Eq. (7.11)]. Then the internal flux distribution within a region can be calculated without difficulty by solving the diffusion equation starting from the given flux and its derivative at the boundary.

The mean flux in a region $\bar{\psi}$ is proportional to the incident currents in that region and expressed as

$$\bar{\psi} = \mathbf{M}^{+}\mathbf{J}^{+}(0) + \mathbf{M}^{-}\mathbf{J}^{-}(1), \tag{7.31}$$

where $\mathbf{J}^{+}(0)$ and $\mathbf{J}^{-}(1)$ are the incident currents in the region and $\mathbf{M}^{+}$ and $\mathbf{M}^{-}$ are matrices which will be called "the mean flux matrices" in subsequent sections.

7.2.7 Extensions to Other One-Dimensional Geometries

Let us consider an infinite cylindrical reactor consisting of many concentric annuli. The response matrices for an annulus can be defined in the same way as those for a slab. The following remarks are necessary.

First, the current density denoted by a column vector $\mathbf{J}^{\pm}$ should be considered as representing the number of neutrons crossing the whole area of the annular surface of a unit length.

Second, the response matrices of an annulus viewed from the inner surface is not equal to those viewed from the outer one, even when the composition is homogeneous.

Third, although neutrons leaving the inner surface of a hollow annulus invariably again strike the inner surface, such additional incident current and the resulting response current should not be included when calculating the response matrices. The reflected or transmitted current is defined as the current of neutrons emerging at the surface for the first time after being introduced to the annulus. The response matrices so defined may be measured by filling the hollow of the annulus with a material perfectly black to neutrons.

Using these remarks, it can be shown that the law of synthesis represented by Eqs. (7.18)–(7.21) is also applicable to the response matrices of annuli. The criticality equation for a cylindrical reactor is obtained by dividing the reactor into inner and outer subassemblies by a dividing interface. Evidently, criticality calculations of spherical reactors can be performed in the same way.

7.3 Calculation of Response Matrices

In this section we shall calculate the response matrices of an infinite homogeneous slab of given thickness based on group diffusion theory.

7.3.1 One-group Diffusion Approximation

Let us first consider the simplest model, in which all the neutrons are assumed to be diffusing in a homogeneous medium with the same energy distribution. In this model, we can separate the energy variable from the space variables. Namely, the neutron flux $\phi(x, E)$ can be expressed as

$$\phi(x, E) = \phi(x) \cdot \psi(E). \tag{7.32}$$

If, in a homogeneous medium, $\psi(E)$ is assumed to be known, the equations for $\phi(x, E)$ can be reduced to energy-independent equations for $\phi(x)$. This model is usually called the "one-group model" and it is often used for the primary analysis.

Let us now inject a steady neutron current of magnitude $J^+(0)$ through the left end of a slab, and observe the internal flux $\phi(x)$ and the response currents $J^-(0)$ and $J^+(a)$. (See Fig. 7.4.)

The equation for $\phi(x)$ is the one-group diffusion equation

$$-D\frac{d^2}{dx^2}\phi(x) + \Sigma\phi(x) = \nu\Sigma_f\phi(x), \tag{7.33}$$

where D, Σ, Σ_f, and ν are, respectively, the diffusion coefficient, the removal cross section, the fission cross section, and the average number of neutrons emitted per fission.

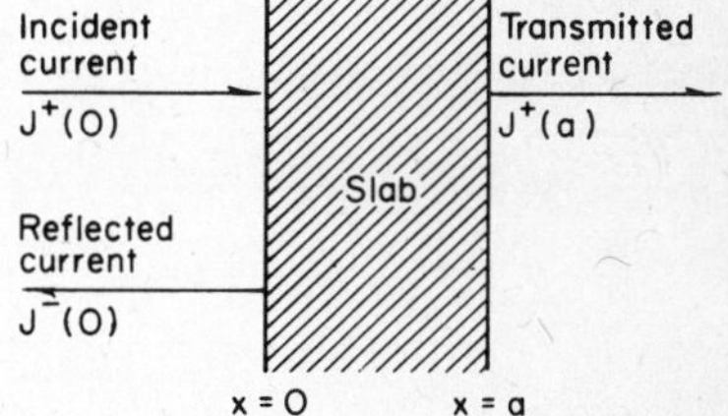

Fig. 7.4 *Response of the slab to the incident current of neutrons.*

According to the diffusion theory, the neutron currents $J^\pm(x)$ are expressed by the flux as

$$J^\pm(x) = \frac{1}{4}\left\{\phi(x) \pm 2D\frac{d}{dx}\phi(x)\right\}. \tag{7.34}$$

Hence the boundary conditions of our present problem are

$$\frac{1}{4}\left\{\phi(x) - 2D\frac{d}{dx}\phi(x)\right\}_{x=0} = J^+(0), \tag{7.35}$$

$$\frac{1}{4}\left\{\phi(x) + 2D\frac{d}{dx}\phi(x)\right\}_{x=a} = 0. \tag{7.36}$$

The reflection coefficient R and the transmission coefficient T can be expressed by $\phi(x)$ as follows:

$$R = \frac{J^-(0)}{J^+(0)} = \frac{\{\phi(x) + 2D(d/dx)\phi(x)\}_{x=0}}{\{\phi(x) - 2D(d/dx)\phi(x)\}_{x=0}}, \tag{7.37}$$

$$T = \frac{J^+(a)}{J^+(0)} = \frac{\{\phi(x) - 2D(d/dx)\phi(x)\}_{x=a}}{\{\phi(x) - 2D(d/dx)\phi(x)\}_{x=0}}. \tag{7.38}$$

Let us solve Eq. (7.33) using the boundary conditions (7.35) and (7.36), and determine R and T explicitly. Equation (7.33) can be rewritten as

$$\frac{d^2}{dx^2}\phi(x) = -\frac{k_\infty - 1}{L^2}\phi(x) \tag{7.39}$$

with $k_\infty = \nu\Sigma_f/\Sigma$ and $L^2 = D/\Sigma$.

(i) *In the case* $k_\infty > 1$, the general solution of (7.39) is

$$\phi(x) = A \sin Bx + C \cos Bx,$$

where

$$B = [(k_\infty - 1)/L^2]^{1/2}.$$

B^2 is the material buckling.

From (7.36) we get

$$C = \frac{2DB \cos Ba + \sin Ba}{2DB \sin Ba - \cos Ba} A,$$

and finally

$$R = \frac{C + 2DAB}{C - 2DAB} = \frac{(1 + 4D^2B^2)\sin Ba}{(1 - 4D^2B^2)\sin Ba + 4DB\cos Ba}. \tag{7.40}$$

Similarly, we get

$$T = \frac{4DB}{(1 - 4D^2B^2)\sin Ba + 4DB\cos Ba}. \tag{7.41}$$

(ii) *In the case* $k_\infty = 1$, the solution of (7.33) is a linear function of x, and we can easily obtain R and T. They are also obtained from (7.40) and (7.41) by a passage to the limit with B tending to zero. We get

$$R = \frac{a}{a + 4D}, \qquad T = \frac{4D}{a + 4D}.$$

(iii) *In the case* $0 \leq k_\infty < 1$, the solution of (7.33) is a linear combination of $\sinh Bx$ and $\cosh Bx$. By a method similar to that of (i), we obtain the results

$$R = \frac{(1 - 4D^2B^2)\sinh Ba}{(1 + 4D^2B^2)\sinh Ba + 4DB\cosh Ba},$$

$$T = \frac{4DB}{(1 + 4D^2B^2)\sinh Ba + 4DB\cosh Ba},$$

where

$$B = [(1 - k_\infty)/L^2]^{1/2}.$$

For a very thick slab ($Ba \gg 1$), we obtain the asymptotic values

$$R \simeq \frac{1 - 2DB}{1 + 2DB},$$

$$T \simeq \frac{8DB}{(1 + 2DB)^2} e^{-Ba}.$$

7.3.2 A Modified Method of Solution

We wish to show another procedure for determining R and T. In fact, this procedure is particularly important for the G-group diffusion model. For the sake of simplicity, we again consider the one-group diffusion model. Furthermore, we assume $k_\infty > 1$. The procedure consists of replacing the exact boundary condition (7.36) by an approximate one and then making an exact correction for the results.

As an approximate boundary condition, let us take the simplest one,

$$\phi(a) = 0. \tag{7.42}$$

Let $\tilde{\phi}(x)$ be the solution of Eq. (7.33) satisfying the boundary conditions (7.35) and (7.42). Then by virtue of condition (7.42), we can conclude that

$$\tilde{\phi}(x) = A' \sin B(a - x). \tag{7.43}$$

Putting $\tilde{\phi}(x)$ into (7.37) and (7.38), we get the approximate coefficients $\tilde{R}$ and $\tilde{T}$ as follows:

$$\tilde{R} = \frac{\sin Ba - 2DB \cos Ba}{\sin Ba + 2DB \cos Ba}, \tag{7.44}$$

$$\tilde{T} = \frac{2DB}{\sin Ba + 2DB \cos Ba}. \tag{7.45}$$

If we set

$$\frac{1}{4}\left\{\tilde{\phi}(x) + 2D \frac{d}{dx} \tilde{\phi}(x)\right\}_{x=a} = W \cdot J^+(0), \tag{7.46}$$

then $\tilde{\phi}(x)$ can be interpreted as the exact solution of (7.33) when the two currents $J^+(0)$ and $W \cdot J^+(0)$ are injected simultaneously into the slab, the former through $x = 0$, the latter through $x = a$. (See Fig. 7.5.) Thus the following expressions are possible:

$$\tilde{R}J^+(0) = RJ^+(0) + TWJ^+(0), \tag{7.47}$$

$$\tilde{T}J^+(0) = TJ^+(0) + RWJ^+(0). \tag{7.48}$$

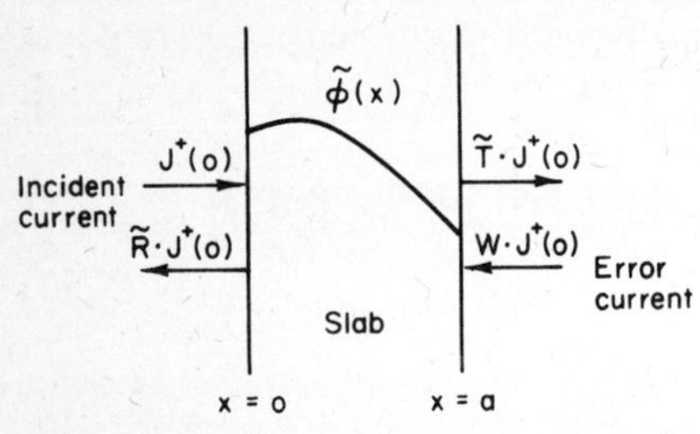

Fig. 7.5 *Relations between error-current and approximate coefficients.*

Hence we get

$$R = (\tilde{R} - \tilde{T}W)(1 - W^2)^{-1}, \tag{7.49}$$

$$T = (\tilde{T} - \tilde{R}W)(1 - W^2)^{-1}. \tag{7.50}$$

From Eqs. (7.46), (7.35), and (7.43) we find

$$W = \frac{\{\tilde{\phi}(x) + 2D(d/dx)\tilde{\phi}(x)\}_{x=a}}{\{\tilde{\phi}(x) - 2D(d/dx)\tilde{\phi}(x)\}_{x=0}} = \frac{-2DB}{\sin Ba + 2DB \cos Ba} = -\tilde{T}.$$

We finally obtain the same results as (7.40) and (7.41). The same applies for the case $k_\infty \leq 1$.

Hereafter we shall call W the error-current coefficient (matrix).

7.3.3 *G*-group Diffusion Approximation

Let us calculate the response matrices based on G-group diffusion theory. We shall follow directly the modified method described previously. The equation for the flux is the G-group diffusion equation

$$
\begin{aligned}
-D_1 \frac{d}{dx^2} \phi_1(x) + \Sigma_1\phi_1(x) &= \nu\Sigma_{fG}\phi_G(x), \\
-D_2 \frac{d}{dx^2} \phi_2(x) + \Sigma_2\phi_2(x) &= \Sigma_{s1}\phi_1(x), \\
&\vdots \\
-D_G \frac{d}{dx^2} \phi_G(x) + \Sigma_G\phi_G(x) &= \Sigma_{s\,G-1}\phi_{G-1}(x),
\end{aligned}
\tag{7.51}
$$

where D_n, Σ_n, Σ_{sn}, and Σ_{fn} are the diffusion coefficient, the removal cross section, the slowing down cross section, and the fission cross section of group n, respectively.*

The neutron current $J_n^{\pm}(x)$ can be expressed by the flux as

$$J_n^{\pm}(x) = \tfrac{1}{4}\{\phi_n(x) \mp d_n \cdot \phi_n'(x)\}, \tag{7.52}$$

where we have set

$$2D_n = d_n.$$

Let us assume that a steady current $J_m^+(0)$ is incident on the surface $x = 0$, and induces the internal flux $\phi_{nm}(x)$ and the response currents $J_{nm}^-(0)$, $J_{nm}^+(a)$ $(n = 1, 2, \ldots, G)$.

We may choose $J_m^+(0)$ as unity; then the elements of the response matrices are

$$R_{nm} = \frac{J_{nm}^-(0)}{J_m^+(0)} = \tfrac{1}{4}\{\phi_{nm}(0) + d_n \cdot \phi_{nm}'(0)\}, \tag{7.53}$$

$$T_{nm} = \frac{J_{nm}^+(a)}{J_m^+(0)} = \tfrac{1}{4}\{\phi_{nm}(a) - d_n \cdot \phi_{nm}'(a)\}. \tag{7.54}$$

Let us solve Eq. (7.51) for $\phi_{nm}(x)$ $(n = 1, 2, \ldots, G)$. From the theory of linear differential equations with constant coefficients, we can express the general solution of (7.51) by the linear combination of $2G$ functions $\{e^{\pm iB_\mu x};\ \mu = 1, 2, \ldots, G\}$. The $\{B_\mu\}$ are the roots of

$$\frac{k_\infty}{(1 + L_1^2B^2)(1 + L_2^2B^2)\cdots(1 + L_G^2B^2)} = 1, \tag{7.55}$$

* For convenience, we have assumed that fissions are induced only by neutrons of group G (generally called the thermal group) and that neutrons of group n are always slowed down to group $n + 1$. These assumptions, however, are not essential to the present discussions.

where

$$L_n{}^2 = \frac{D_n}{\Sigma_n},$$

$$k_\infty = \left(\frac{\nu\Sigma_{fG}}{\Sigma_G}\right)\left(\frac{\Sigma_{s1}}{\Sigma_1}\right)\cdots\left(\frac{\Sigma_{s\,G-1}}{\Sigma_{G-1}}\right).$$

If we impose the exact boundary conditions

$$\tfrac{1}{4}\{\phi_{nm}(0) - d_n \cdot \phi'_{nm}(0)\} = \delta_{nm} \tag{7.56}$$

and

$$J^-_{nm}(a) = \tfrac{1}{4}\{\phi_{nm}(a) + d_n \cdot \phi'_{nm}(a)\} = 0, \tag{7.57}$$

then we have to solve $2G$ simultaneous linear equations for the coefficients of linear combinations. But, as before, by the use of the approximate boundary condition,

$$\phi_{nm}(a) = 0 \qquad (n = 1, 2, \ldots, G) \tag{7.58}$$

instead of (7.57), we can reduce the problem to that of solving G simultaneous equations. Thus we shall first obtain the approximate response matrices $\tilde{\mathbf{R}}$ and $\tilde{\mathbf{T}}$. Next, we define the error-current matrix $\mathbf{W} = (W_{nm})$ by

$$W_{nm} = \tfrac{1}{4}\{\tilde{\phi}_{nm}(a) + d_n \cdot \tilde{\phi}'_{nm}(a)\} = -\tilde{T}_{nm},$$

where $\tilde{\phi}_{nm}(x)$ is the solution of (7.51) satisfying conditions (7.56) and (7.58). Then we find the matrix equations analogous to (7.47) and (7.48):

$$\tilde{\mathbf{R}} = \mathbf{R} + \mathbf{T} \cdot \mathbf{W},$$

$$\tilde{\mathbf{T}} = \mathbf{T} + \mathbf{R} \cdot \mathbf{W}.$$

Using the relation $\mathbf{W} = -\tilde{\mathbf{T}}$, we finally get

$$\mathbf{R} = [\tilde{\mathbf{R}} + \tilde{\mathbf{T}}^2][\mathbf{E} - \tilde{\mathbf{T}}^2]^{-1}, \tag{7.59}$$

$$\mathbf{T} = [\tilde{\mathbf{T}} + \tilde{\mathbf{R}} \cdot \tilde{\mathbf{T}}][\mathbf{E} - \tilde{\mathbf{T}}^2]^{-1}. \tag{7.60}$$

In Appendix B, the reader will find the formulas for calculating the elements of $\tilde{\mathbf{R}}$ and $\tilde{\mathbf{T}}$ for the case of two-group diffusion approximation.

7.3.4 Numerical Examples

In this section we shall give some numerical values of the response matrices of various homogeneous slabs. The study was based on two-group diffusion approximation. Various materials commonly used as reactor components were selected.

Table 7.1 *Group Constants of Various Materials*

Materials	τ (cm^2)[a]	L^2 (cm^2)[b]	D_1 (cm)	D_2 (cm)
H_2O	33.1	7.45	1.42	0.164
D_2O	138	8060	1.31	0.62
Be	102	484	0.687	0.60
BeO	112	1030	0.645	0.75
C	420	2250	1.19	0.778
Al	15,900	396	5.52	5.52
Fe	182	1.61	0.345	0.345
Zr	3560	130	1.22	0.99

[a] $\tau = D_1/\Sigma_1$.
[b] $L^2 = D_2/\Sigma_2$.

The materials and the group constants which were used are listed in Table 7.1. Elements of the reflection matrices of slabs with infinite thickness are given in Table 7.2. Table 7.3 gives response matrices for various slabs of 20-cm thickness. Figures 7.6–7.11 show the elements of the response matrices as a function of thickness for various materials.

As an example for the reactor core, we shall give the response matrices of the homogenized core of a research reactor* (Toshiba Training Reactor) at various thicknesses (cf. Tables 7.4 and 7.5).

* TTR is a research reactor similar to the Oak Ridge Research Reactor.

Table 7.2 *Reflection Matrix for Infinite Thickness*

Materials	R_{11}	R_{21}	R_{22}
H_2O	0.3390	0.1897	0.7854
D_2O	0.6356	0.3178	0.9728
Be	0.7605	0.1557	0.8966
BeO	0.7830	0.1559	0.9107
C	0.7919	0.1407	0.9364
Al	0.8389	0.0141	0.2864
Fe	0.9027	0.0054	0.2955
Zr	0.9215	0.0107	0.7041

Table 7.3 *Response Matrices of Slabs of 20-cm Thickness*

Materials	R_{11}	R_{21}	R_{22}	T_{11}	T_{21}	T_{22}
H_2O	0.3387	0.1897	0.7854	0.0274	0.0140	0.00025
D_2O	0.6228	0.1732	0.8877	0.1103	0.0922	0.1092
Be	0.7543	0.1171	0.8637	0.0588	0.0545	0.0909
BeO	0.7759	0.1055	0.8520	0.0595	0.0520	0.1202
C	0.7459	0.0616	0.8569	0.1543	0.0361	0.1296
Al	0.4679	0.00555	0.2507	0.5183	0.00457	0.3397
Fe	0.8937	0.00534	0.2955	0.0439	0.00039	0.13×10^{-6}
Zr	0.7957	0.00622	0.6933	0.1909	0.00343	0.0886

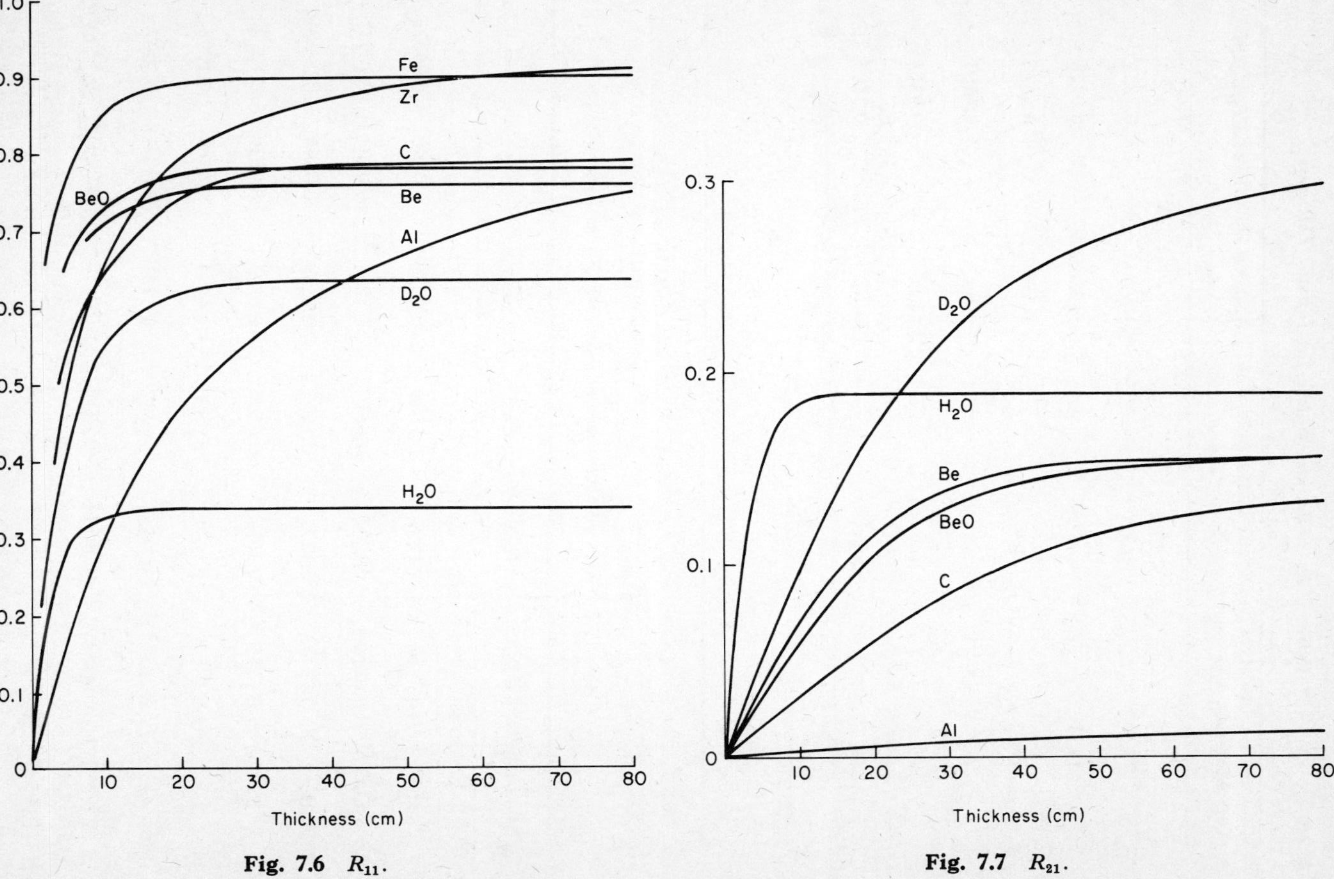

Fig. 7.6 R_{11}.

Fig. 7.7 R_{21}.

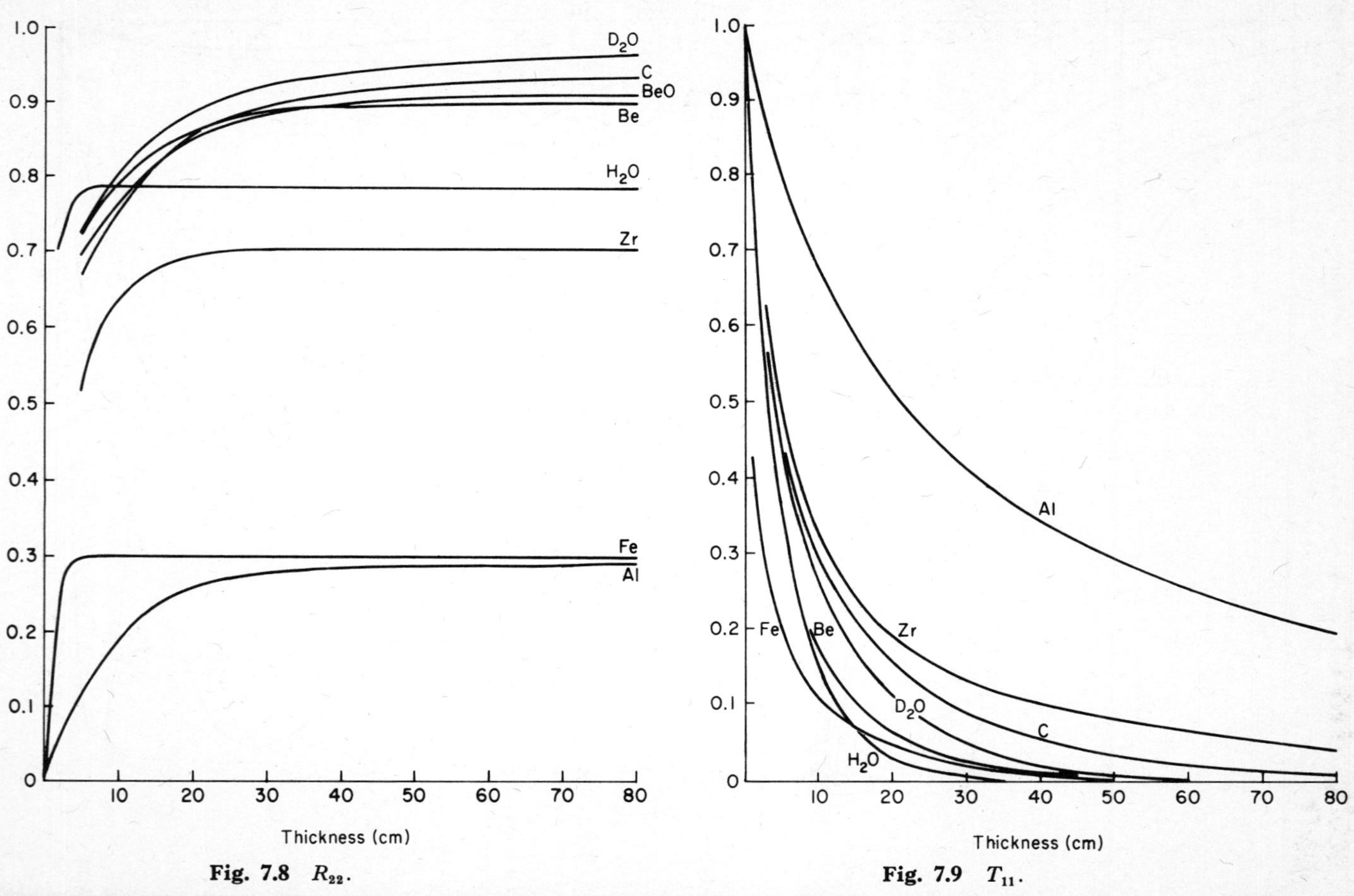

Fig. 7.8 R_{22}.

Fig. 7.9 T_{11}.

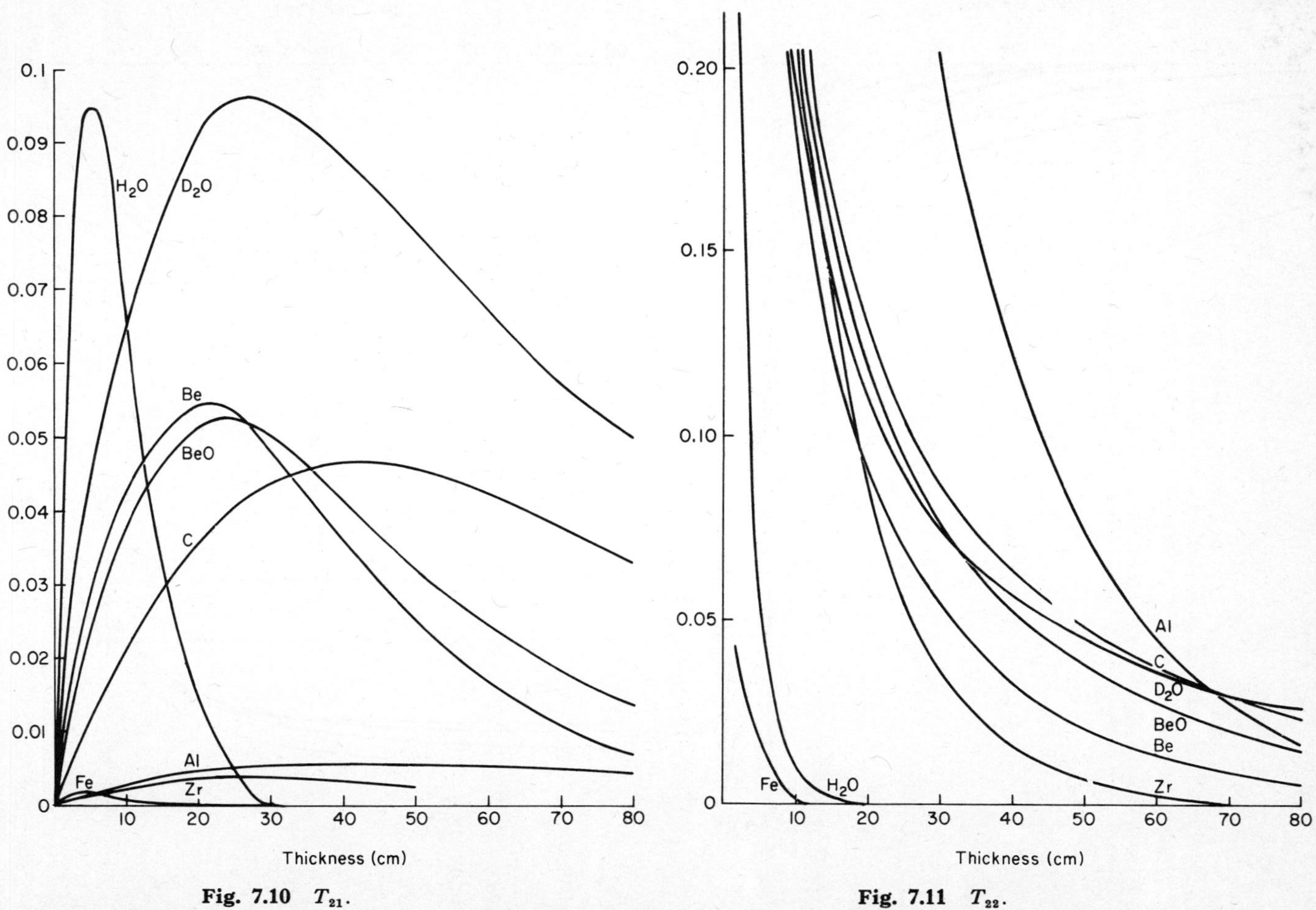

Fig. 7.10 T_{21}.

Fig. 7.11 T_{22}.

Table 7.4 *Reflection Matrix of a Homogenized Core*

Thickness (cm)	R_{11}	R_{12}	R_{21}	R_{22}
5	0.52816	0.33798	0.04990	0.60381
10	0.68152	0.44957	0.06638	0.61756
15	0.76656	0.50788	0.07499	0.62347
20	0.83033	0.55132	0.08140	0.62784
15.576[a]	0.77454	0.51333	0.07579	0.62402

[a] Half the critical thickness.

Table 7.5 *Transmission Matrix of a Homogenized Core*

Thickness (cm)	T_{11}	T_{12}	T_{21}	T_{22}
5	0.38342	0.20077	0.02964	0.06573
10	0.25080	0.16626	0.02455	0.01984
15	0.20338	0.13815	0.02040	0.01410
20	0.18713	0.12738	0.01881	0.01282
15.576[a]	0.20030	0.13613	0.02010	0.01384

[a] Half the critical thickness.

7.4 Criticality Calculation

In this section we shall discuss various methods of criticality calculation. Although we again consider the slab reactor, the methods are almost directly applicable to other one-dimensional geometries.

Let us assume that the reactor is composed of $(M + M')$ homogeneous slabs whose response matrices $\mathbf{R}_m$, $\mathbf{T}_m$ are already known (cf. Fig. 7.12).

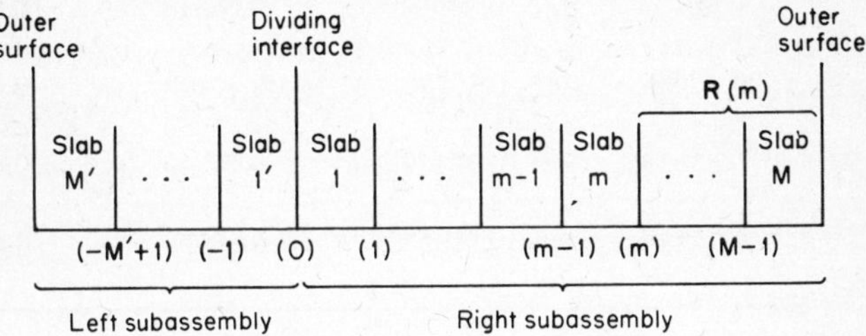

Fig. 7.12 *Division of a slab reactor.*

7.4.1 Analytical Method

By successive use of the "law of synthesis," we can synthesize the matrices $\mathbf{R}_r^+$ and $\mathbf{R}_l^-$ (the reflection matrices of the right and the left subassemblies viewed from the dividing interface).

Since the component slabs are assumed to be homogeneous, the law of synthesis which is represented by Eqs. (7.18) and (7.19) becomes simpler. Let $\mathbf{R}(m)$ be the reflection matrix of the combined region composed of slab $m+1, \ldots,$ slab M. Then we find the following recurrence formula:

$$\mathbf{R}(m-1) = \mathbf{R}_m + \mathbf{T}_m\mathbf{R}(m) \cdot \{\mathbf{E} - \mathbf{R}_m\mathbf{R}(m)\}^{-1} \cdot \mathbf{T}_m, \quad m = M-1, \ldots, 1, \tag{7.61}$$

with

$$\mathbf{R}(M-1) = \mathbf{R}_M.$$

The reflection matrix $\mathbf{R}_r^+$ of the right subassembly is equal to $\mathbf{R}(0)$. From the recurrence relation, we similarly obtain $\mathbf{R}_l^-$.

The eigenvalue problem (7.24) can be solved analytically, if G (number of groups) is not very large. Thus we can perform the criticality calculations at the expense of only $M + M' - 2$ matrix inversions and one matrix eigenvalue problem of order G.

7.4.2 Iterative Method

The iterative method is also applicable to our eigenvalue problem. Let us inject an arbitrary current $\mathbf{J}_0^+(0)$ into the right subassembly through the dividing surface (0), and trace the life cycle which we have already defined in Section 7.2. After one cycle we get a new current $\tilde{\mathbf{J}}_0^+(0)$ which appears from the left subassembly. To trace the next cycle, we define a scalar λ_1 and a current $\mathbf{J}_1^+(0)$ as follows:

$$\lambda_1 = \frac{\| \tilde{\mathbf{J}}_0^+(0) \|}{\| \mathbf{J}_0^+(0) \|}, \tag{7.62}$$

$$\mathbf{J}_1^+(0) = \frac{1}{\lambda_1} \tilde{\mathbf{J}}_0^+(0). \tag{7.63}$$

We trace the next cycle by injecting $\mathbf{J}_1^+(0)$ into the right subassembly. By iterating this procedure, we get a convergent sequence

$$\{\lambda_1, \mathbf{J}_1^+(0)\}, \quad \{\lambda_2, \mathbf{J}_2^+(0)\}, \ldots, \rightarrow \{\lambda, \mathbf{J}^+(0)\}$$

whose limit is the solution of our eigenvalue problem.

This is the outline of the iterative method. The actual process involves the following calculations.

(i) *For the right subassembly,*

$$\mathbf{J}_t^+(m) = \mathbf{T}_m \mathbf{J}_t^+(m-1) + \mathbf{R}_m \mathbf{J}_{t-1}^-(m), \qquad m = 1, 2, \ldots, M-1, \tag{7.64}$$

$$\mathbf{J}_t^-(m-1) = \mathbf{T}_m \mathbf{J}_t^-(m) + \mathbf{R}_m \mathbf{J}_t^+(m-1), \qquad m = M, M-1, \ldots, 1, \tag{7.65}$$

with $\mathbf{J}_t^-(M) = 0$.

(ii) *For the left subassembly,*

$$\mathbf{J}_t^-(m) = \mathbf{T}_m\mathbf{J}_t^-(m+1)+\mathbf{R}_m\mathbf{J}_{t-1}^+(m), \qquad m = -1, -2, \ldots, -M'+1, \tag{7.66}$$

$$\mathbf{J}_t^+(m+1) = \mathbf{T}_m\mathbf{J}_t^+(m)+\mathbf{R}_m\mathbf{J}_t^-(m+1), \qquad m = -M', \ldots, -1, \tag{7.67}$$

with $\mathbf{J}_t^+(-M') = 0$.

The linear extrapolation technique or some other method can be employed to accelerate the convergence.

Besides the eigenvalue λ and the eigencurrent $\mathbf{J}^+(0)$, we can obtain the currents $\mathbf{J}^\pm(m)$ at each interface. These currents are useful in calculating the internal flux or mean flux within each component slab.

7.4.3 Determination of Internal Flux and Mean Flux

Since we already know the neutron currents at each interface, it is easy to obtain the internal flux and the mean flux within each slab. For example, let us consider the slab m with respect to which we already know the following quantities:

$$\mathbf{J}^\pm(m-1), \quad \mathbf{J}^\pm(m)$$

(cf. Fig. 7.13). Then the internal flux $\boldsymbol{\phi}(x)$ can be obtained by solving the diffusion equation with the initial conditions

$$\boldsymbol{\phi}(0) = 2[\mathbf{J}^+(m-1) + \mathbf{J}^-(m-1)] \tag{7.68}$$

and

$$\boldsymbol{\phi}'(0) = \mathbf{D}^{-1}[\mathbf{J}^-(m-1) - \mathbf{J}^+(m-1)], \tag{7.69}$$

where $\mathbf{D}$ is the diagonal matrix composed of diffusion coefficients.

By using the mean flux matrix* $\mathbf{M}$, we can also obtain the mean flux as

$$\boldsymbol{\psi} = \mathbf{M}[\mathbf{J}^+(m-1) + \mathbf{J}^-(m)]. \tag{7.70}$$

* The elements of $\mathbf{M} = \{M_{nl}\}$ are obtained by

$$M_{nl} = \frac{1}{a}\int_0^a \phi_{nl}(x)\,dx,$$

where $\phi_{nl}(x)$ is the neutron flux of group n induced by a unit incident current of group l at $x = 0$. This calculation can be performed in the course of calculating $\mathbf{R}$ and $\mathbf{T}$.

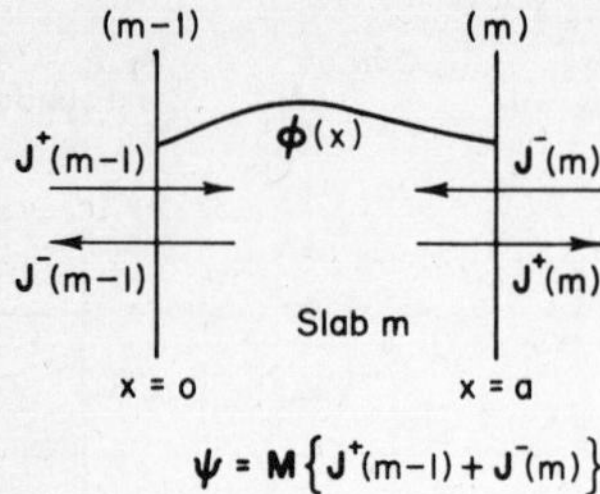

Fig. 7.13 *Internal flux and mean flux.*

7.4.4 Criticality Calculation Using an Analog Computer

Let us consider the use of an analog computer for the criticality calculation of a reactor.

We can simulate the response of a black box by a combination of potentiometers and adders in an analog computer. In the 2-group approximation, for example, the black box is simulated by 4 adders and 16 potentiometers as shown in Fig. 7.14. Here the response matrices are represented by the setting of the potentiometers.

The entire reactor is then simulated by a serial combination of these unit circuits, each of which represents the component black box. The criticality calculation can be performed by connecting the left and the

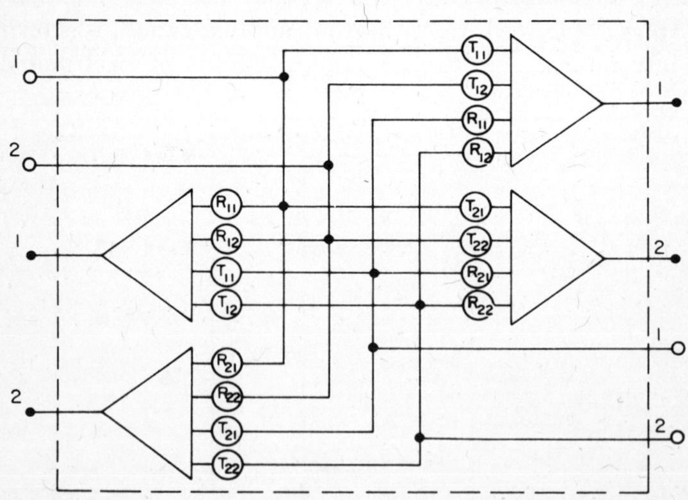

Fig. 7.14 *Simulator for one-dimensional region.* ○ *input terminal,* ● *output terminal,* ◯ *potentiometer,* △ *adder.*

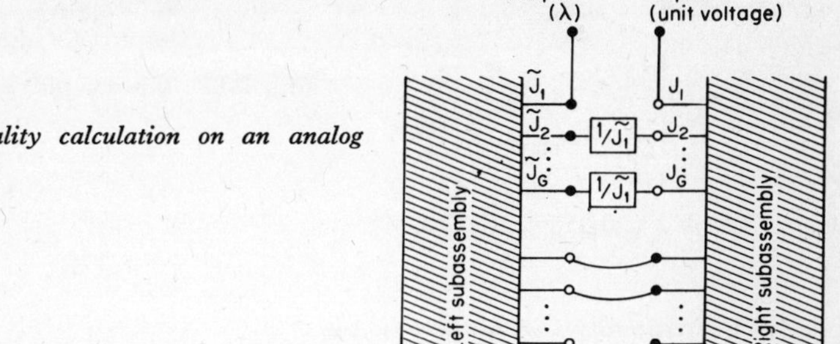

Fig. 7.15 *Criticality calculation on an analog computer.*

right subassemblies as shown in Fig. 7.15, and by injecting J_1 into the right subassembly. In view of the definition of the neutron life cycle, the connection should be made so that the relation

$$\frac{\tilde{J}_n}{\tilde{J}_1} = \frac{J_n}{J_1}$$

holds for $n = 2, 3, \ldots, G$.

The eigenvalue λ is given by the ratio $\tilde{J}_1/J_1$. This method is quite analogous to the iterative method described previously. If the reactor has a plane of symmetry, it is sufficient to simulate the reactor only for the right (or left) subassembly, and we can observe $\sqrt{\lambda}$ as the output voltage. An example, with the two-group approximation, is shown in Fig. 7.16. An integrator is used to avoid the instability of electronic circuits.

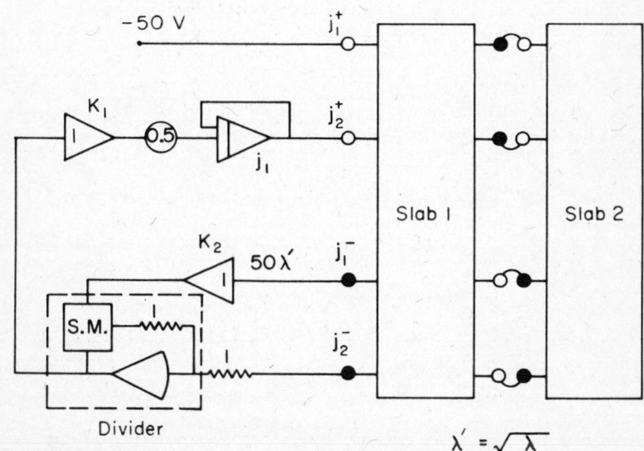

Fig. 7.16 *Block diagram for criticality calculation with two-group approximation.*

By using the analog computer, we can obtain λ within an accuracy of $0.1 \sim 0.5\%$. The accuracy depends on the number of slabs (black boxes) involved.

7.5 Examples of Criticality Calculations

In concluding this chapter, we wish to emphasize that the response matrix method gives us the exact solution of the one-dimensional diffusion equations because neither assumptions nor approximations were used in the formulations described so far. Secondly, it should be mentioned that by this method the computation time is remarkably reduced com-

Table 7.6 *Comparison with Finite Difference Method*[a]

Method:		Finite difference method			Response matrix method
Code:		SIC			MERMAID-1A
Number of grid points (slabs):		25	50	95	(3)
Relative computation time:[b]	$t = 10$	60	100	200	
	$t = 4$	24	40	80	1
Multiplication factor:[b]	$t = 10$	1.001295	1.000531	1.000149	
	$t = 4$	1.001695	1.000548	1.000566	1.000000
Flux at interface:[c]	$\phi_1(0)$	1.00000	1.00000	1.00000	1.000000
	$\phi_2(0)$	0.23034	0.23033	0.23032	0.230315
	$\phi_1(1)$	0.54287	0.54179	0.54125	0.541026
	$\phi_2(1)$	0.20002	0.19844	0.19753	0.197161
	$\phi_1(2)$	0.16590	0.17060	0.17177	0.172158
	$\phi_2(2)$	0.23431	0.22635	0.22327	0.222110
Mean core flux:	$\bar{\phi}_1$			0.84862	0.848535
	$\bar{\phi}_2$	0.20513	0.20461	0.20444	0.204361

[a] Geometry: bisymmetrical slab reactor. Composition: region 1, TTR core, region 2, graphite reflector, region 3, water reflector.

[b] t represents the number of source iterations.

[c] $\phi_1(n)$ represents fast flux at the interface between region n and region $n + 1$ and ϕ_2 represents thermal flux.

Table 7.7 *Results of Analog Computation*[a]

Reflector materials	Analog computation (TOSAC II[b])			Digital computation (USSC)
	Case 1	Case 2	Case 3	
H_2O	0.993	0.992	0.992	0.9924
D_2O	1.018	1.019	1.018	1.0192
Be	1.058	1.060	1.060	1.0596
C	1.016	1.017	1.016	1.0172
Al	0.960	0.960	0.960	0.9597
Fe	1.054	1.055	1.055	1.0547
Zr	1.007	1.007	1.008	1.009
BeO	1.060	1.060	1.059	1.0598

[a] Geometry: bisymmetrical slab reactor. Composition: region 1, TTR core (18.325 cm); region 2, reflector (10 cm).

[b] Made by Toshiba.

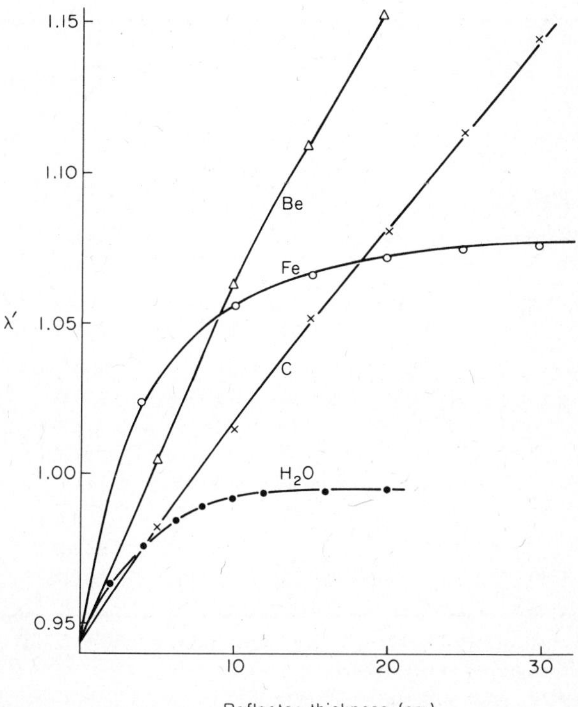

Fig. 7.17 *Effect of reflector thickness on criticality. TTR core (36.65 cm).*

pared to the time required for the conventional method (the finite-difference method).

As an illustration, we shall show the numerical results of the criticality calculations of a research reactor (TTR). Two codes, MERMAID-1A* (response matrix method) and SIC† (finite difference method), were used for comparison. Since the results of the SIC calculations are dependent on the number of grid points and the number of source iterations, several combinations of these numbers were tested (cf. Table 7.6).

Finally we shall show the results of the analog computations performed for the parametric study of the reflector effects. The criticality factors of a reflected reactor are listed in Table 7.7 for various reflector materials. The results of the digital computations are listed for comparison. We can see the effect of the reflector thickness on the criticality factor in Fig. 7.17.

* Programmed for the USSC-90 by one of the authors.
† Programmed for the USSC-90 by Mr. Tokizawa of NAIG Co.

CHAPTER EIGHT *TWO-DIMENSIONAL PROBLEM*

8.1 Introduction

In this chapter we shall discuss, along the line of group-diffusion theory, the application of the response matrix method to the criticality calculations of two-dimensional reactors.

Because of its simplicity and wide applicability for multidimensional problems, the few-group diffusion theory continues to be the primary nuclear design tool. The most widely used technique for the solution of two-dimensional diffusion equations is the finite difference method. According to this method, one must solve a set of large scale, simultaneous linear equations, the solution of which require the use of the matrix iterative method. For example, if we divide the reactor into 50×50 mesh points, then we must solve 2500 simultaneous linear equations many times in order to obtain the eigenvalue (the criticality factor) and the corresponding eigenfunction (the neutron flux).

Thus the two-dimensional diffusion code based on the finite difference method requires several minutes of computer time per problem, and often requires more than 30 minutes on an IBM-7090 class computer.

This situation greatly motivates us to try to apply the response matrix method to the two-dimensional criticality calculations.

In applying the response matrix method to the two-dimensional problems, there is a difficulty in that the response of a two-dimensional rod depends not only on the total intensity but also on the intensity distribution of the incident current at the incidence surface. Thus, in order to simplify the dependence of the response on the intensity distribution, it is necessary to make an efficient approximation for the intensity distribution of the incident current.

The simplest approximation may be that the intensity is uniformly distributed on the incidence surface. But it was found that this approximation always overestimates the leakage of neutrons from the reactor and does not give reasonable results, even when the size of the component rods is very small. The next approximation that should be examined will be that the intensity distribution is a linear function (of the coordinate on the surface of incidence) which involves an adjustable parameter. It was found, from numerical comparisons with the finite difference method, that this approximation is good enough even when the size of the component rods is fairly large.

In the next section, we shall give details of this latter approximation together with the definitions of two-dimensional response matrices. Calculations of two-dimensional response matrices can be reduced to those of the one-dimensional problems by expanding the incident current and the internal flux in terms of a complete set of orthogonal functions. The mathematical formulation will be found in Section 8.3.

For criticality calculations, an iterative method can be constructed on a principle similar to that used in the one-dimensional problems. Details of the iterative method together with the method of determining the aforementioned adjustable parameter will be described in Section 8.4.

Along these lines, a two-dimensional diffusion code MERMAID-2 was programmed for the IBM-7090. Numerical comparisons with the finite difference method proved that the response matrix method is efficient for two-dimensional problems also, again reducing computation time appreciably. Computation time for a typical problem is less than 1 minute with an error, in the criticality factor, of about 0.1%. MERMAID-2 was extended to MERMAID-2B which performs burnup calculations. In this code, the criticality search is performed by regulating the control rods at each time step of fuel depletion. An example of the burnup calculation will be found in Section 8.6.

8.2 Definitions of Two-Dimensional Response Matrices

In this section we shall define two-dimensional response matrices based on G-group diffusion theory.

Let us observe the response of an infinite rectangular rod to a steady current of the mth energy group, $I_m{}^+(x, 0)$, incident on the surface $y = 0$ (see Fig. 8.1). In Fig. 8.1 the output currents of the nth group, due to $I_m{}^+(x, 0)$, are indicated by $I_{nm}^-(x, 0)$, $I_{nm}^+(x, b)$, $J_{nm}^-(-a/2, y)$, and $J_{nm}^+(a/2, y)$, respectively.

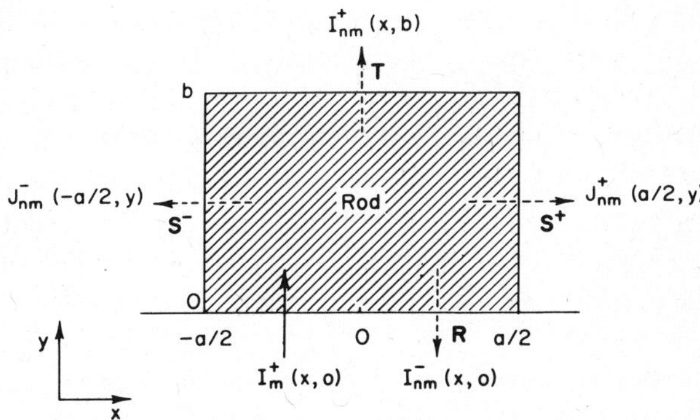

Fig. 8.1 *Response of an infinite rectangular rod.*

If the functional form of the incident current is specified, then the functional forms of the output currents are uniquely determined by the composition and the size of the rod. Thus, it will be quite natural to characterize the response of the rod by the ratios of the integrals of the output currents to that of the incident current. For example, the reflection may be characterized by the following ratio:

$$R_{ynm} = \frac{\int_{-a/2}^{a/2} I_{nm}^-(x, 0)\, dx}{\int_{-a/2}^{a/2} I_m{}^+(x, 0)\, dx} = \frac{\int_{-a/2}^{a/2} I_{nm}^-(x, 0)\, dx}{I_m{}^+(0)}. \tag{8.1}$$

We use the subscript y to denote the direction of incidence. Thus we can define the reflection matrix of the rod with respect to the y-directional incidence. It is

$$\mathbf{R}_y = \{R_{ynm}\} \qquad (n, m = 1, 2, \ldots, G). \tag{8.2}$$

Similarly, the transmission matrix $\mathbf{T}_y$ and the transverse transmission matrices $\mathbf{S}_y^{\pm}$ can be defined, the elements of which are, respectively,

$$T_{ynm} = \frac{\int_{-a/2}^{a/2} I_{nm}^{+}(x, b)\, dx}{I_m^{+}(0)}, \tag{8.3}$$

$$S_{ynm}^{+} = \frac{\int_0^b J_{nm}^{+}(a/2, y)\, dy}{I_m^{+}(0)}, \tag{8.4}$$

$$S_{ynm}^{-} = \frac{\int_0^b J_{nm}^{-}(-a/2, y)\, dy}{I_m^{+}(0)}. \tag{8.5}$$

It should be noted here that the matrices defined above are dependent not only on the composition and the size of the rod but also on the specified form of the incident current. For example, the ratio S_{ynm}^{-} will be smaller in the case of Fig. 8.2a than in the case of Fig. 8.2b. So it does not seem realistic to fix the functional form of every incident current to a particular form. On the other hand, however, it is not practical to treat the functional form of the individual current in full detail. In other words, we have to forge an efficient approximation to the functional form of the incident current.

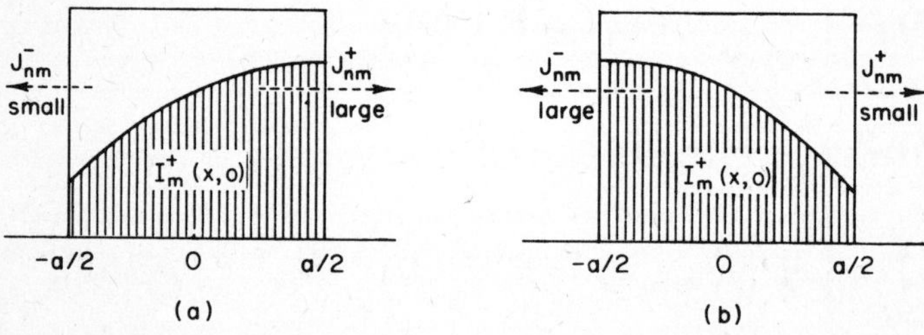

Fig. 8.2 *Effect of intensity distribution of incident current.*

A reasonable approximation in this case is the linear approximation. Let us assume that every incident current can be approximated by a linear function of the surface coordinate, for example,

$$I_m^{+}(x, 0) = \left\{1 + \alpha_m^{y}\left(\frac{2}{a}\right)x\right\} I_m^{+}(0), \tag{8.6}$$

where α_m^{y} is an adjustable parameter representing the average obliquity of $I_m^+(x, 0)$. In vector form, our assumption is expressed as

$$\mathbf{I}^+(x, 0) = \left\{\mathbf{E} + \left(\frac{2}{a}\right) x \boldsymbol{\alpha}^y\right\} \mathbf{I}^+(0) \tag{8.7}$$

with

$$\mathbf{I}^+(0) = \begin{bmatrix} I_1^+(0) \\ I_2^+(0) \\ \vdots \\ I_G^+(0) \end{bmatrix}, \quad \boldsymbol{\alpha}^y = \begin{bmatrix} \alpha_1^y & & & 0 \\ & \alpha_2^y & & \\ & & \ddots & \\ 0 & & & \alpha_G^y \end{bmatrix}. \tag{8.8}$$

Since the incident current is of the form (8.7), it can be shown that $\mathbf{R}_y$ and $\mathbf{T}_y$ are independent of $\boldsymbol{\alpha}^y$ and that $\mathbf{S}_y^{\pm}$ can be represented as

$$\mathbf{S}_y^{\pm} = \mathbf{S}_y \pm \mathbf{S}_y' \boldsymbol{\alpha}^y, \tag{8.9}$$

where $\mathbf{S}_y$ and $\mathbf{S}_y'$ are matrices dependent only on the composition and the size of the rod.

Thus the two-dimensional response of the rod to the incident current, given by (8.7), can be expressed by the following equations:

$$\mathbf{I}^-(0) = \int_{-a/2}^{a/2} \mathbf{I}^-(x, 0)\, dx = \mathbf{R}_y \mathbf{I}^+(0), \tag{8.10}$$

$$\mathbf{I}^+(b) = \int_{-a/2}^{a/2} \mathbf{I}^+(x, b)\, dx = \mathbf{T}_y \mathbf{I}^+(0), \tag{8.11}$$

$$\mathbf{J}^+\left(\frac{a}{2}\right) = \int_0^b \mathbf{J}^+\left(\frac{a}{2}, y\right) dy = \mathbf{S}_y^+ \mathbf{I}^+(0) = (\mathbf{S}_y + \mathbf{S}_y' \boldsymbol{\alpha}^y) \mathbf{I}^+(0), \tag{8.12}$$

$$\mathbf{J}^-\left(-\frac{a}{2}\right) = \int_0^b \mathbf{J}^-\left(-\frac{a}{2}, y\right) dy = \mathbf{S}_y^- \mathbf{I}^+(0) = (\mathbf{S}_y - \mathbf{S}_y' \boldsymbol{\alpha}^y) \mathbf{I}^+(0). \tag{8.13}$$

The response matrices and the equations corresponding to the x-directional incidence are obtained similarly by exchanging x, a, and $\mathbf{I}$ with y, b, and $\mathbf{J}$. The response of the rod to the simultaneous incidence at the four surfaces is readily described using the "principle of superposition."

Now we wish to mention that the matrix $\boldsymbol{\alpha}^y$ should be evaluated for each y-directional incident current $\mathbf{I}$, and $\boldsymbol{\alpha}^x$ for each x-directional incident current **J**. These matrices can be determined consistently in the course of the criticality calculation. The details will be found in Section 8.4.

Based on G-group diffusion theory, we shall give in the next section the mathematical formulation for the calculation of the response matrices $\mathbf{R}_x$, $\mathbf{R}_y$, $\mathbf{T}_x$, $\mathbf{T}_y$, $\mathbf{S}_x$, $\mathbf{S}_y$, $\mathbf{S}_x'$, and $\mathbf{S}_y'$ which are unique to the rod.

8.3 Calculation of Response Matrices

Let $\phi_{nm}(x, y)$ be the internal flux of the nth group due to the incident current $I_m^+(x, 0)$. Then $\phi_{nm}(x, y)$ $(n = 1, 2, \ldots, G)$ satisfy the coupled equations (two-dimensional group diffusion equation)

$$\begin{aligned}
-D_1\left(\frac{\partial^2}{\partial x^2} + \frac{\partial^2}{\partial y^2}\right)\phi_{1m} + \Sigma_1\phi_{1m} &= \nu\Sigma_{fG}\phi_{Gm}, \\
-D_2\left(\frac{\partial^2}{\partial x^2} + \frac{\partial^2}{\partial y^2}\right)\phi_{2m} + \Sigma_2\phi_{2m} &= \Sigma_{s1}\phi_{1m}, \qquad (8.14)\\
&\;\;\vdots \\
-D_G\left(\frac{\partial^2}{\partial x^2} + \frac{\partial^2}{\partial y^2}\right)\phi_{Gm} + \Sigma_G\phi_{Gm} &= \Sigma_{s\,G-1}\phi_{G-1\,m},
\end{aligned}$$

and the boundary conditions

$$\phi_{nm} + d_n \frac{\partial}{\partial x}\phi_{nm} = 0 \qquad \text{at} \quad x = \frac{a}{2}, \tag{8.15}$$

$$\phi_{nm} - d_n \frac{\partial}{\partial x}\phi_{nm} = 0 \qquad \text{at} \quad x = -\frac{a}{2}, \tag{8.16}$$

$$\phi_{nm} + d_n \frac{\partial}{\partial y}\phi_{nm} = 0 \qquad \text{at} \quad y = b, \tag{8.17}$$

$$\phi_{nm} - d_n \frac{\partial}{\partial y}\phi_{nm} = 4\,\delta_{nm}I_m^+(x, 0) \qquad \text{at} \quad y = 0, \tag{8.18}$$

$$n = 1, 2, \ldots, G,$$

where we have set $2D_n = d_n$.

In principle, we can solve Eq. (8.14) using the above boundary conditions. By inserting the flux-current relations

$$I^{\pm}(x, y) = \frac{1}{4}\left\{\phi(x, y) \mp 2D \frac{\partial}{\partial y} \phi(x, y)\right\}, \tag{8.19}$$

$$J^{\pm}(x, y) = \frac{1}{4}\left\{\phi(x, y) \mp 2D \frac{\partial}{\partial x} \phi(x, y)\right\} \tag{8.20}$$

into Eqs. (8.1)–(8.5), we obtain the expressions

$$R_{ynm} = \frac{\int_{-a/2}^{a/2} dx\ \frac{1}{4}\{\phi_{nm} + d_n(\partial/\partial y)\phi_{nm}\}_{y=0}}{I_m^{+}(0)}, \tag{8.21}$$

$$T_{ynm} = \frac{\int_{-a/2}^{a/2} dx\ \frac{1}{4}\{\phi_{nm} - d_n(\partial/\partial y)\phi_{nm}\}_{y=b}}{I_m^{+}(0)}, \tag{8.22}$$

$$S_{ynm}^{+} = \frac{\int_0^b dy\ \frac{1}{4}\{\phi_{nm} - d_n(\partial/\partial x)\phi_{nm}\}_{x=a/2}}{I_m^{+}(0)}, \tag{8.23}$$

$$S_{ynm}^{-} = \frac{\int_0^b dy\ \frac{1}{4}\{\phi_{nm} + d_n(\partial/\partial x)\phi_{nm}\}_{x=-a/2}}{I_m^{+}(0)}. \tag{8.24}$$

In practice, it is not an easy task to solve Eq. (8.14) using the exact boundary conditions. The difficulty is that d_n depends on n. So we shall again use the approximate boundary conditions

$$\phi_{nm} + d \frac{\partial}{\partial x} \phi_{nm} = 0 \qquad \text{at} \quad x = \frac{a}{2}, \tag{8.25}$$

$$\phi_{nm} - d \frac{\partial}{\partial x} \phi_{nm} = 0 \qquad \text{at} \quad x = -\frac{a}{2}, \tag{8.26}$$

$$n = 1, 2, \ldots, G,$$

instead of (8.15) and (8.16), where d is a constant which we can select. (In Section 7.3 we have treated the case $d = 0$.) To be concrete, let us set d equal to $2D_G$.

8.3.1 A Complete Set of Orthogonal Functions

We can find a complete set of orthogonal functions in x by solving the equation

$$\frac{d^2}{dx^2} \varphi(x) + B^2\varphi(x) = 0 \tag{8.27}$$

using the boundary conditions

$$\varphi + d\frac{d}{dx}\varphi = 0 \quad \text{at} \quad x = \frac{a}{2}, \tag{8.28}$$

$$\varphi - d\frac{d}{dx}\varphi = 0 \quad \text{at} \quad x = -\frac{a}{2}, \tag{8.29}$$

where B^2 is the eigenvalue.

After some simple arithmetic, it is shown that the set is composed of even functions

$$\varphi_k{}^+(x) = \cos B_k{}^+x \tag{8.30}$$

and odd functions

$$\varphi_k{}^-(x) = \sin B_k{}^-x, \tag{8.31}$$

where

$$B_k{}^\pm = \frac{2}{a}\,\xi_k{}^\pm. \tag{8.32}$$

The $\xi_k{}^+$ and $\xi_k{}^-$ are, respectively, the kth positive root of the transcendental equations

$$\cos \xi_k{}^+ - \left(\frac{2d}{a}\right)\xi_k{}^+ \sin \xi_k{}^+ = 0, \tag{8.33}$$

$$\sin \xi_k{}^- + \left(\frac{2d}{a}\right)\xi_k{}^- \cos \xi_k{}^- = 0. \tag{8.34}$$

8.3.2 Expansions in Terms of $\{\varphi_k{}^\pm(x)\}$

In the calculation of response matrices, the assumption that the integrated intensity of the incident current is unity is not a restriction.

Now let us expand the incident current and the internal fluxes in terms of $\{\varphi_k{}^\pm(x)\}$:

$$I_m{}^+(x, 0) = 1 + \alpha_m{}^y\left(\frac{2}{a}\right)x = \sum_k \{A_{mk}^+\varphi_k{}^+(x) + A_{mk}^-\varphi_k{}^-(x)\}, \tag{8.35}$$

$$\phi_{nm}(x, y) = \sum_k \{A_{mk}^+ f_{nm}^{+(k)}(y)\varphi_k{}^+(x) + A_{mk}^- f_{nm}^{-(k)}(y)\varphi_k{}^-(x)\}. \tag{8.36}$$

By substituting (8.36) into (8.14), (8.17), and (8.18), we find the corresponding one-dimensional equations and the boundary conditions for $f_{nm}^{\pm(k)}(y)$ $(k = 1, 2, \ldots)$.

They are

$$-D_1 \frac{d^2}{dy^2} f_{1m}^{\pm(k)} + \{\Sigma_1 + D_1(B_k^{\pm})^2\} f_{1m}^{\pm(k)} = \nu\Sigma_{fG} f_{Gm}^{\pm(k)},$$

$$-D_2 \frac{d^2}{dy^2} f_{2m}^{\pm(k)} + \{\Sigma_2 + D_2(B_k^{\pm})^2\} f_{2m}^{\pm(k)} = \Sigma_{s1} f_{1m}^{\pm(k)}, \tag{8.37}$$

$$\vdots$$

$$-D_G \frac{d^2}{dy^2} f_{Gm}^{\pm(k)} + \{\Sigma_G + D_G(B_k^{\pm})^2\} f_{Gm}^{\pm(k)} = \Sigma_{s\,G-1} f_{G-1\,m}^{\pm(k)},$$

and

$$f_{nm}^{\pm(k)} + d_n \frac{d}{dy} f_{nm}^{\pm(k)} = 0 \qquad \text{at} \quad y = b, \tag{8.38}$$

$$f_{nm}^{\pm(k)} - d_n \frac{d}{dy} f_{nm}^{\pm(k)} = 4\,\delta_{nm} \qquad \text{at} \quad y = 0, \tag{8.39}$$

$$n = 1, 2, \ldots, G.$$

Thus, our problem is now reduced to the one-dimensional problem which we have already solved in Section 7.3 [cf. Eqs. (7.51), (7.56), and (7.57)]. In Eq. (8.37), Σ_n is replaced by $\Sigma_n + D_n(B_k^{\pm})^2$.

8.3.3 Expressions for Response Matrices

The equations can be solved for each k and we can calculate the following quantities:

$$R_{nm}^{(k)} = \frac{1}{4}\left\{f_{nm}^{+(k)} + d_n \frac{d}{dy} f_{nm}^{+(k)}\right\}_{y=0}, \tag{8.40}$$

$$T_{nm}^{(k)} = \frac{1}{4}\left\{f_{nm}^{+(k)} - d_n \frac{d}{dy} f_{nm}^{+(k)}\right\}_{y=b}, \tag{8.41}$$

$$M_{nm}^{\pm(k)} = \frac{1}{b}\int_0^b dy\, f_{nm}^{\pm(k)}. \tag{8.42}$$

By substituting (8.35) and (8.36) into (8.21)–(8.24) and performing all the integrations, we finally obtain the expressions

$$R_{ynm} = \frac{\sum_k C_{mk} R_{nm}^{(k)}}{\sum_k C_{mk}}, \tag{8.43}$$

$$T_{ynm} = \frac{\sum_k C_{mk} T_{nm}^{(k)}}{\sum_k C_{mk}}, \tag{8.44}$$

$$S_{ynm}^{\pm} = S_{ynm} \pm \alpha_m{}^y S'_{ynm}, \tag{8.45}$$

where

$$C_{mk} = A_{mk}^{+} \int_{-a/2}^{a/2} \varphi_k{}^{+}(x)\, dx = \frac{2a \sin^2 \xi_k{}^{+}}{\xi_k{}^{+}(\xi_k{}^{+} + \frac{1}{2} \sin 2\xi_k{}^{+})}, \tag{8.46}$$

$$S_{ynm} = \frac{b}{4}\left(1 + \frac{d_n}{d}\right) \frac{\sum_k D_{mk}^{+} M_{nm}^{+(k)}}{\sum_k C_{mk}}, \tag{8.47}$$

$$S'_{ynm} = \frac{b}{4}\left(1 + \frac{d_n}{d}\right) \frac{\sum_k D_{mk}^{-} M_{nm}^{-(k)}}{\sum_k C_{mk}}, \tag{8.48}$$

$$D_{mk}^{+} = A_{mk}^{+} \varphi_k{}^{+}\left(\frac{a}{2}\right) = \frac{\sin 2\xi_k{}^{+}}{\xi_k{}^{+} + \frac{1}{2} \sin 2\xi_k{}^{+}}, \tag{8.49}$$

$$D_{mk}^{-} = \frac{A_{mk}^{-}}{\alpha_m} \varphi_k{}^{-}\left(\frac{a}{2}\right) = \frac{2 \sin^2 \xi_k{}^{-} - \xi_k{}^{-} \sin 2\xi_k{}^{-}}{\xi_k{}^{-}(\xi_k{}^{-} + \frac{1}{2} \sin 2\xi_k{}^{-})}. \tag{8.50}$$

In practice, the summation over k must be handled by a finite number of calculations using a suitable convergence criterion.

If one desires to obtain the mean flux due to the incident current, it is convenient to calculate the "mean flux matrix" F defined by

$$F_{ynm} = \frac{\int_{-a/2}^{a/2} dx \int_0^b dy\, \phi_{nm}(x, y)}{abI_m{}^{+}(0)} = \frac{a^{-1} \sum_k C_{mk} M_{nm}^{+(k)}}{\sum_k C_{mk}}. \tag{8.51}$$

Response matrices $\mathbf{R}_x$, $\mathbf{T}_x$, $\mathbf{S}_x$, $\mathbf{S}_x'$, and $\mathbf{F}_x$ corresponding to the x-directional incidence can be calculated similarly.

8.3.4 A Method for Improving the Accuracy of the Response Matrices

Because of the approximate boundary conditions (8.25) and (8.26), the response matrices derived above are not exact (except for the case $D_1 = D_2 = \cdots = D_G$). The exact matrices can be obtained by introducing the error-current matrices and relating them to the approximate matrices as has been described in Section 7.3 [cf. Eqs. (7.47)–(7.50)].

In the two-dimensional case, however, the procedure is somewhat complicated because we have to take into account the intensity distributions not only of the incident currents but also of the error currents. A detailed discussion of this problem will be found in the original paper [29].

In this section we shall describe a practical method for improving the accuracy. Let us first define the error-current matrices by the integrations of the error currents over the surfaces $x = a/2$ and $x = -a/2$:

$$W_{ynm}^{+} = \int_0^b dy \frac{1}{4} \left\{ \tilde{\phi}_{nm} + d_n \frac{\partial}{\partial x} \tilde{\phi}_{nm} \right\}_{x=a/2} \Big/ I_m{}^{+}(0), \tag{8.52}$$

$$W_{ynm}^{-} = \int_0^b dy \frac{1}{4} \left\{ \tilde{\phi}_{nm} - d_n \frac{\partial}{\partial x} \tilde{\phi}_{nm} \right\}_{x=-a/2} \Big/ I_m{}^{+}(0), \tag{8.53}$$

where $\tilde{\phi}_{nm}(x, y)$ is the solution subject to the approximate boundary conditions (8.25) and (8.26). By a simple calculation, we find

$$W_{ynm}^{\pm} = W_{ynm} \pm \alpha_m{}^y W_{ynm}' = \left(\frac{d - d_n}{d + d_n} \right) (\tilde{S}_{ynm} \pm \alpha_m{}^y \tilde{S}_{ynm}'). \tag{8.54}$$

Hereafter we shall use the notations $\tilde{\mathbf{R}}_y$, $\tilde{\mathbf{R}}_x$, $\tilde{\mathbf{T}}_y$, $\tilde{\mathbf{T}}_x$, $\tilde{\mathbf{S}}_y$, $\tilde{\mathbf{S}}_x$, $\tilde{\mathbf{S}}_y'$, and $\tilde{\mathbf{S}}_x'$ for the approximate response matrices calculated from $\tilde{\phi}_{nm}(x, y)$ [using Eqs. (8.43)–(8.50)], to distinguish them from the exact (or improved) ones $\mathbf{R}_y$, $\mathbf{R}_x$, etc. which we are now going to obtain.

Now let us find relations among these matrices, assuming that the intensity distribution of the error currents (on surfaces $x = \pm a/2$) is

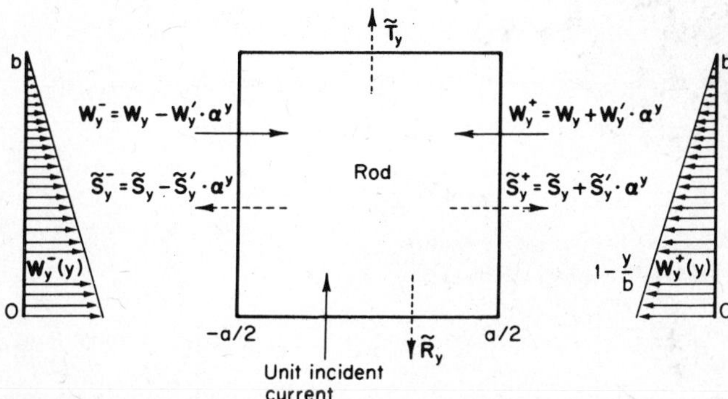

Fig. 8.3 *Relations between exact and approximate response matrices.*

given by* (see Fig. 8.3)

$$1 - \frac{y}{b}. \tag{8.55}$$

Utilizing the relations $\mathbf{W}_y^- + \mathbf{W}_y^+ = 2\mathbf{W}_y$ and $\mathbf{S}_x^\pm = \mathbf{S}_x \mp \mathbf{S}_x'$,* we easily obtain the following relations:

$$\tilde{\mathbf{R}}_y = \mathbf{R}_y + 2(\mathbf{S}_x + \mathbf{S}_x')\mathbf{W}_y, \tag{8.56}$$

$$\tilde{\mathbf{T}}_y = \mathbf{T}_y + 2(\mathbf{S}_x - \mathbf{S}_x')\mathbf{W}_y, \tag{8.57}$$

$$\tilde{\mathbf{S}}_y = \mathbf{S}_y + (\mathbf{R}_x + \mathbf{T}_x)\mathbf{W}_y, \tag{8.58}$$

$$\tilde{\mathbf{S}}_y' = \mathbf{S}_y' + (\mathbf{R}_x - \mathbf{T}_x)\mathbf{W}_y'. \tag{8.59}$$

Similarly, we find the following relations with respect to the x-directional incidence:

$$\tilde{\mathbf{R}}_x = \mathbf{R}_x + 2(\mathbf{S}_y + \mathbf{S}_y')\mathbf{W}_x, \tag{8.60}$$

$$\tilde{\mathbf{T}}_x = \mathbf{T}_x + 2(\mathbf{S}_y - \mathbf{S}_y')\mathbf{W}_x, \tag{8.61}$$

$$\tilde{\mathbf{S}}_x = \mathbf{S}_x + (\mathbf{R}_y + \mathbf{T}_y)\mathbf{W}_x, \tag{8.62}$$

$$\tilde{\mathbf{S}}_x' = \mathbf{S}_x' + (\mathbf{R}_y - \mathbf{T}_y)\mathbf{W}_x'. \tag{8.63}$$

From (8.56)–(8.63) we finally obtain

$$\mathbf{S}_y = [\tilde{\mathbf{S}}_y - (\tilde{\mathbf{R}}_x + \tilde{\mathbf{T}}_x) \cdot \mathbf{W}_y] \cdot (\mathbf{E} - 4\mathbf{W}_x \cdot \mathbf{W}_y)^{-1}, \tag{8.64}$$

$$\mathbf{S}_y' = [\tilde{\mathbf{S}}_y' - (\tilde{\mathbf{R}}_x - \tilde{\mathbf{T}}_x) \cdot \mathbf{W}_y'] \cdot (\mathbf{E} - 4\mathbf{W}_x \cdot \mathbf{W}_y')^{-1}, \tag{8.65}$$

$$\mathbf{S}_x = [\tilde{\mathbf{S}}_x - (\tilde{\mathbf{R}}_y + \tilde{\mathbf{T}}_y) \cdot \mathbf{W}_x] \cdot (\mathbf{E} - 4\mathbf{W}_y \cdot \mathbf{W}_x)^{-1}, \tag{8.66}$$

$$\mathbf{S}_x' = [\tilde{\mathbf{S}}_x' - (\tilde{\mathbf{R}}_y - \tilde{\mathbf{T}}_y) \cdot \mathbf{W}_x'] \cdot (\mathbf{E} - 4\mathbf{W}_y \cdot \mathbf{W}_x')^{-1}, \tag{8.67}$$

$$\mathbf{R}_y = \tilde{\mathbf{R}}_y - 2(\mathbf{S}_x + \mathbf{S}_x') \cdot \mathbf{W}_y, \tag{8.68}$$

$$\mathbf{T}_y = \tilde{\mathbf{T}}_y - 2(\mathbf{S}_x - \mathbf{S}_x') \cdot \mathbf{W}_y, \tag{8.69}$$

$$\mathbf{R}_x = \tilde{\mathbf{R}}_x - 2(\mathbf{S}_y + \mathbf{S}_y') \cdot \mathbf{W}_x, \tag{8.70}$$

$$\mathbf{T}_x = \tilde{\mathbf{T}}_x - 2(\mathbf{S}_y - \mathbf{S}_y') \cdot \mathbf{W}_x. \tag{8.71}$$

* This corresponds to the assumption that $\alpha_m^x = -1$ $(m = 1, 2, \ldots, G)$ for the error currents.

8.3.5 Relation between Response Matrices

In view of the neutron conservation in the rod, there must be a relation between the elements of the response matrices. By considering the conservation of neutrons of group n induced by a unit incident current of group m, we can find the following relation:

$$R_{nm} + T_{nm} + 2S_{nm} + F_{nm}ab\Sigma_n$$
$$= \delta_{nm} + \delta_{n1}F_{Gm}ab(\nu\Sigma_f)_G + F_{n-1\ m}ab\Sigma_{s\ n-1} \quad (\text{for} \quad n, m = 1, 2, \ldots, G) \tag{8.72}$$

with

$$F_{0m} = 0.$$

From a practical point of view this is a very important relation, because the criticality of a reactor is strongly related to the conservation of neutrons. In practice, (8.72) can be used for final adjustment of the truncation or round-off errors committed in the course of calculations of response matrices.

8.3.6 Examples of Response Matrices

Elements of the response matrices of homogeneous square rods of various sizes are given in Table 8.2. For a square rod, $\mathbf{R}_x$, $\mathbf{T}_x$, $\mathbf{S}_x$, and $\mathbf{S}_x'$ evidently coincide with $\mathbf{R}_y$, $\mathbf{T}_y$, $\mathbf{S}_y$, and $\mathbf{S}_y'$, respectively. The group constants used for the calculation are given in Table 8.1. The computation time for a square rod is about 1 second on an IBM-7090.

Table 8.1 Core Group Constants

k_∞	1.612
p	0.8102
τ	299 cm^2
L^2	77.8 cm^2
D_1	0.9 cm
D_2	0.9 cm
B_{z1}^2	0.4959×10^{-3} cm^2
B_{z2}^2	0.4959×10^{-3} cm^2

Table 8.2 *Response Matrices of Square Rods*

	Size:	a^a	$a/2$	$a/4$	$a/10$	$a/20$
R	R_{11}	0.76442	0.64976	0.50183	0.25152	0.04656
	R_{12}	0.36093	0.24077	0.13814	0.05327	0.02403
	R_{21}	0.03824	0.02551	0.01464	0.005644	0.002546
	R_{22}	0.63022	0.56023	0.45046	0.23172	0.03762
T	T_{11}	0.01596	0.020823	0.04117	0.10259	0.1790
	T_{12}	0.02841	0.03244	0.03429	0.02357	0.01501
	T_{21}	0.003010	0.003437	0.003633	0.002498	0.001590
	T_{22}	0.005400	0.008762	0.02842	0.09383	0.1734
S	S_{11}	0.09557	0.13965	0.20765	0.31223	0.3815
	S_{12}	0.08635	0.08228	0.06551	0.03457	0.01871
	S_{21}	0.009150	0.008717	0.006941	0.003663	0.001983
	S_{22}	0.06346	0.10906	0.18329	0.29938	0.3746
S′	S'_{11}	0.05744	0.08141	0.10556	0.12062	0.1102
	S'_{12}	0.03015	0.02339	0.01377	0.004706	0.001574
	S'_{21}	0.003194	0.002478	0.001459	0.000499	0.0001669
	S'_{22}	0.04623	0.07272	0.10044	0.11887	0.1097

[a] $a = 68.745$ cm.

8.4 Criticality Calculation

8.4.1 Criticality Equation

Using the response matrices of the component rods, we can perform the criticality calculation for a two-dimensional reactor. For this purpose we divide the reactor into upper and lower subassemblies by a dividing surface. Let the upper subassembly be composed of $L \times K$ component rods and the lower subassembly of $L \times K'$ component rods as shown in Fig. 8.4.

Suppose incident currents $\mathbf{I}_1^+ \sim \mathbf{I}_L^+$ are introduced simultaneously into the upper subassembly through the dividing surface. Then, after one cycle, the reflected currents $\tilde{\mathbf{I}}_1^+ \sim \tilde{\mathbf{I}}_L^+$ will appear from the lower

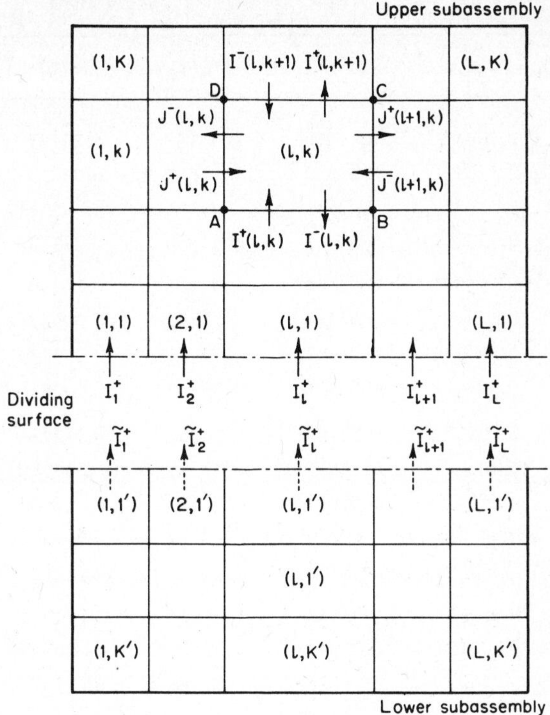

Fig. 8.4 *Division of a two-dimensional reactor.*

subassembly. The criticality equation corresponding to Eq. (7.22) can be written as

$$\tilde{\mathbf{I}}_l^+ = \lambda \mathbf{I}_l^+ \qquad (l = 1, 2, \ldots, L). \tag{8.73}$$

Iterative methods for solving (8.73) may be constructed using the principle described in Section 7.4 [cf. Eqs. (7.62)–(7.67)]. But we think it more pertinent to explain a method for determining $\boldsymbol{\alpha}$ before describing the details of the iterative method.

8.4.2 Determination of $\boldsymbol{\alpha}$

Let us consider the rod (l, k) and determine the diagonal matrix $\boldsymbol{\alpha}^y(l, k)$ which represents the average obliquity of current $\mathbf{I}^+(l, k)$ on the surface AB (cf. Fig. 8.4).

Assuming that the intensity distribution of $\mathbf{I}^{+}(l, k)$ has the same form as the flux distribution on the surface AB, we can express the nth component of $\boldsymbol{\alpha}^{y}(l, k)$ by

$$\alpha_n{}^y(l, k) \simeq \frac{[\phi_n(B) - \phi_n(A)]}{[\phi_n(B) + \phi_n(A)]}. \tag{8.74}$$

Further, if we adopt the assumption that both $\phi_n(A)$ and $\phi_n(B)$ can be approximated by the mean flux on the surfaces AD and BC, respectively, Eq. (8.74) can be rewritten as

$$\alpha_n{}^y(l, k) \simeq \frac{[J_n{}^+(l+1, k) + J_n{}^-(l+1, k)] - [J_n{}^+(l, k) + J_n{}^-(l, k)]}{[J_n{}^+(l+1, k) + J_n{}^-(l+1, k)] + [J_n{}^+(l, k) + J_n{}^-(l, k)]}. \tag{8.75}$$

According to this formulation, the average obliquity of $\mathbf{I}^{-}(l, k+1)$ on DC is also represented by the matrix $\boldsymbol{\alpha}^{y}(l, k)$ given above.

Similarly, $\boldsymbol{\alpha}^{x}(l, k)$, which represents the average obliquity of $\mathbf{J}^{+}(l, k)$ and $\mathbf{J}^{-}(l+1, k)$, is determined by

$$\alpha_n{}^x(l, k) \simeq \frac{[I_n{}^+(l, k+1) + I_n{}^-(l, k+1)] - [I_n{}^+(l, k) + I_n{}^-(l, k)]}{[I_n{}^+(l, k+1) + I_n{}^-(l, k+1)] + [I_n{}^+(l, k) + I_n{}^-(l, k)]}. \tag{8.76}$$

8.4.3 Equations for Partial Currents

The basic equations used for the iterative procedure are

$$\begin{aligned}\mathbf{I}^{+}(l, k+1) = {} & \mathbf{T}_y(l, k)\mathbf{I}^{+}(l, k) + \mathbf{R}_y(l, k)\mathbf{I}^{-}(l, k+1) \\ & + [\mathbf{S}_x(l, k) + \mathbf{S}_x{}'(l, k)\boldsymbol{\alpha}^x(l, k)][\mathbf{J}^{+}(l, k) + \mathbf{J}^{-}(l+1, k)],\end{aligned} \tag{8.77}$$

$$\begin{aligned}\mathbf{I}^{-}(l, k) = {} & \mathbf{R}_y(l, k)\mathbf{I}^{+}(l, k) + \mathbf{T}_y(l, k)\mathbf{I}^{-}(l, k+1) \\ & + [\mathbf{S}_x(l, k) - \mathbf{S}_x{}'(l, k)\boldsymbol{\alpha}^x(l, k)][\mathbf{J}^{+}(l, k) + \mathbf{J}^{-}(l+1, k)],\end{aligned} \tag{8.78}$$

$$\begin{aligned}\mathbf{J}^{-}(l, k) = {} & \mathbf{T}_x(l, k)\mathbf{J}^{-}(l+1, k) + \mathbf{R}_x(l, k)\mathbf{J}^{+}(l, k) \\ & + [\mathbf{S}_y(l, k) - \mathbf{S}_y{}'(l, k)\boldsymbol{\alpha}^y(l, k)][\mathbf{I}^{-}(l, k+1) + \mathbf{I}^{+}(l, k)],\end{aligned} \tag{8.79}$$

$$\begin{aligned}\mathbf{J}^{+}(l+1, k) = {} & \mathbf{T}_x(l, k)\mathbf{J}^{+}(l, k) + \mathbf{R}_x(l, k)\mathbf{J}^{-}(l+1, k) \\ & + [\mathbf{S}_y(l, k) + \mathbf{S}_y{}'(l, k)\boldsymbol{\alpha}^y(l, k)][\mathbf{I}^{-}(l, k+1) + \mathbf{I}^{+}(l, k)].\end{aligned} \tag{8.80}$$

The boundary conditions at the outer surfaces can be introduced quite naturally. Let us consider, for example, the left boundary. If it is a free surface, there should be no currents entering the reactor. On the other hand, if it is a boundary of symmetry, net currents should vanish at this surface. Thus we have the boundary conditions at the left boundary:

$$\mathbf{I}^{+}(1, k) = 0 \qquad \text{(free boundary)}, \tag{8.81}$$

$$\mathbf{I}^{+}(1, k) = \mathbf{I}^{-}(1, k) \qquad \text{(boundary of symmetry)}. \tag{8.82}$$

Conditions at the other boundaries are obtained similarly.

8.4.4 An Iterative Scheme for Solving the Criticality Equation

With arbitrary initial currents $\mathbf{I}^{+}_{(0)}(l, 1)$ $(l = 1\text{–}L)$ introduced into the upper subassembly, we can begin an iterative procedure based on Eqs. (8.77)–(8.80). A scheme of iteration is explained in Fig. 8.5, where

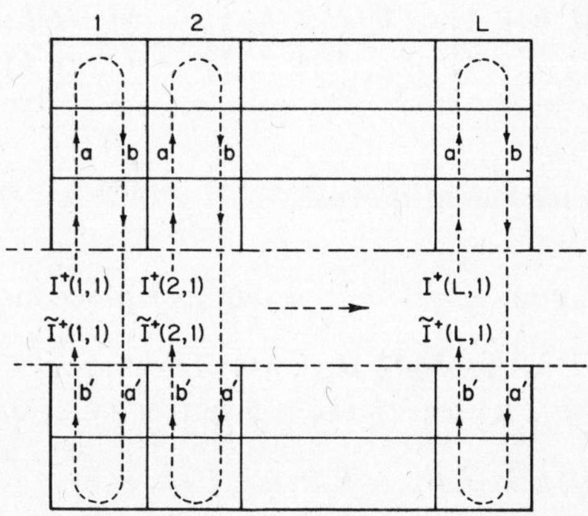

Fig. 8.5 *A scheme of iteration.*

sweep-*a* and sweep-*b* are alternately performed for the upper and lower subassemblies. Sweep-*a* and sweep-*b* are shown schematically in Fig. 8.6, where the currents indicated by dashed lines are computed from the currents indicated by solid lines. In each sweep, the matrices $\boldsymbol{\alpha}^{y}$ and $\boldsymbol{\alpha}^{x}$ can be determined by (8.75) and (8.76) from the newest currents available.

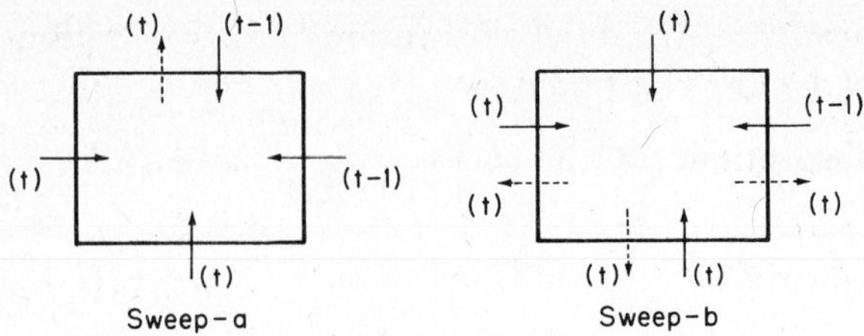

Fig. 8.6 *Sweep-a and sweep-b. Iteration index t.*

At the end of the tth cycle, $\lambda_{(t)}$ may be calculated from

$$\lambda_{(t)} = \left[\frac{\sum_{l=1}^{L} \tilde{I}_1^{\,+}(l, 1)}{\sum_{l=1}^{L} I_1^{\,+}(l, 1)} \right]_{(t)}. \tag{8.83}$$

The next cycle begins with the new currents $\mathbf{I}^+_{(t+1)}(l, 1)$ given by

$$\mathbf{I}^+_{(t+1)}(l, 1) = \frac{\tilde{\mathbf{I}}^+_{(t)}(l, 1)}{\lambda_{(t)}} \qquad (l = 1, 2, \ldots, L). \tag{8.84}$$

These iterations are continued until

$$|\, \lambda_{(t+1)} - \lambda_{(t)} \,| < \varepsilon \tag{8.85}$$

is satisfied for a given convergence criterion ε. An extrapolation method can be applied to accelerate the convergence of the iterations. Let $\mathbf{X}^*_{(t)}$ be the currents computed from Eqs. (8.77)–(8.80); then the accelerated currents are given by

$$\mathbf{X}_{(t)} = \mathbf{X}^*_{(t)} + \omega(\mathbf{X}^*_{(t)} - \mathbf{X}_{(t-1)}), \tag{8.86}$$

where ω is the relaxation factor. Numerical experiments show that the best value for ω lies between 0.2 and 0.4 in most cases. Generally speaking, ω becomes small as the number of component rods increases.

8.4.5 Mean Flux and Neutron Balance

After convergence of the iterative procedure, we obtain the eigencurrents at all the interfaces. From these currents we can determine the mean flux within each component rod by using the matrices $\mathbf{F}_y$ and $\mathbf{F}_x$ defined by (8.51):

$$\boldsymbol{\Psi}(l, k) = \mathbf{F}_x[\mathbf{J}^+(l, k) + \mathbf{J}^-(l+1, k)] + \mathbf{F}_y[\mathbf{I}^+(l, k) + \mathbf{I}^-(l, k+1)]. \tag{8.87}$$

From the mean flux the total absorptions and generations of neutrons in the rod (l, k) are computed:

$$\text{absorptions in rod } (l, k) = ab \sum_{n=1}^{G} \Sigma_{an}\psi_n(l, k), \tag{8.88}$$

$$\text{generations in rod } (l, k) = ab \sum_{n=1}^{G} (\nu\Sigma_f)_n\psi_n(l, k). \tag{8.89}$$

The overall neutron balance in the reactor is then represented by the summation of the above quantities over all the component rods and the total number of neutrons leaking out of the reactor. Evidently, the leakage from the reactor can be obtained by a simple summation of the partial currents over the free surfaces of the reactor. Thus the effective multiplication factor based on the neutron balance will be given by

$$k_{\text{eff}} = \frac{[\text{Generation}]}{[\text{Absorption}] + [\text{Leakage}]}. \tag{8.90}$$

8.5 Examples of Criticality Calculations

Based on the method so far described, a two-dimensional diffusion code MERMAID-2* was programmed for the IBM-7090. According to two-group diffusion theory, this code first calculates the response matrices of the homogeneous rods which constitute the reactor. The computation time for a rod is about 2 seconds (for a square rod it is about 1 second). It is also possible, at the user's option, to input the matrices that were prepared by experiments or by other refined theories.

The criticality calculations are then performed using the iterative method described in the previous section. After convergence of the iteration, the following quantities are obtained as outputs: λ, k_{eff}, mean flux in each rod, and partial currents at each interface.

To prove the efficiency of the response matrix method for two-dimensional problems, several calculations were performed and comparisons with the conventional method were made. Two-dimensional diffusion codes PDQ[7] and 2V-MISA,† which are based on the finite difference method, were used for comparison.

* Programmed by the authors.

† Programmed by K. Aoyama, I. Komatsu, and K. Aoki.

8.5.1 Calculations for a Homogeneous Bare Reactor

For a given set of group constants, we can determine the critical size of a bare square reactor. For the group constants given in Table 8.1, the critical size is 137.5×137.5 cm. A quadrant of this critical reactor was treated both by the conventional method and the response matrix method. The number of grid points or component rods* was increased to ascertain the effects on the accuracy of the multiplication factor or on computation time (cf. Table 8.3).

Table 8.3 *Calculation for a Homogeneous Bare Reactor*

Method	Code	N^a	$k_{eff}{}^b$	C^c	Computation time on IBM-7090 (sec)
Finite difference	2V-MISA	5×5	1.00203	0.051	
		10×10	1.00057	0.057	150
		15×15	1.00019	0.043	
	PDQ	20×20	1.00015	0.061	270
		40×40	1.000028	0.045	600
Response matrix	MERMAID-2	2×2	1.00224	0.0090	4
		4×4	1.00047	0.0075	8
		6×6	1.00020	0.0072	11
		8×8	1.00010	0.0064	29
		10×10	1.000045	0.0045	51
		12×12	1.000017	0.0024	114
		14×14	1.000005	0.0010	176

[a] N is the total number of grid points or component rods.
[b] Exact $k_{eff} = 1.000000$.
[c] $C = N \cdot (k_{eff} - 1)$.

It should be remarked that the value of C is much smaller for the response matrix method and it decreases with increasing N, whereas C is almost constant for the finite difference method.

The required number of iterations for convergence was 111, in the case of 14×14 rods.

* Whose response matrices are listed in Table 8.2.

8.5.2 Calculations for a Simple Reactor with Reflector

Similar comparisons were made for a reactor that has a reflector. The results are shown in Table 8.4. It is worth noting that the size of the component rods can be fairly large. For example, in the case of 3×3 rods it is as large as 22×22 cm, but the multiplication factor is obtained with good accuracy.

Table 8.4 *Calculation for a Simple Reflected Reactor*

Method	Code	N	k_{eff}	Computation time on IBM-7090 (sec)
Finite difference	2V-MISA	13×13	1.0024	53
		34×34	1.0000	1022
Response matrix	MERMAID-2	3×3	1.0013	4
		6×6	0.99907	17
		9×9	0.99942	46
		12×12	0.99956	95

8.5.3 Calculation for a Multiregion Reactor

The result of the criticality calculation for TTR is shown in Table 8.5. Again, a quadrant was calculated. A fuel assembly (size 7.73×7.73 cm) was treated as a component rod for the calculation using MERMAID-2. The number of iterations required for convergence was 48.

Table 8.5 *Calculation for Toshiba Training Reactor*

Method	Code	N	k_{eff}	Computation time on IBM-7090
Finite difference	PDQ	51×55	1.0725	23 min
Response matrix	MERMAID-2	7×8	1.0711	30 sec

8.6 Application to the Burnup Calculation

In burnup calculations, one must take the "change of group constants due to fuel depletion" into account. For this purpose one usually divides the time into a number of "time steps" and assumes that the group constants and the flux distributions are kept unchanged in each time step. The fuel depletion is computed at the end of each time step, and the group constants for the fuel regions are recalculated at the beginning of the next time step. Naturally, it is necessary that the reactor always be kept in the critical condition. The criticality of a reactor is usually attained by regulating the position of the control rods. Accordingly, many criticality calculations must be performed at each time step in order to know the position of the control rods and the corresponding neutron flux. This process is usually called the "criticality search."

All these computation processes are repeated until the end of the core life. So it will be understood that the burnup calculation for a two- or three-dimensional reactor requires much computation time.

MERMAID-2 was further extended to MERMAID-2B* for the burnup calculations of two-dimensional reactors. In this code, the group constants of the fuel regions are fitted through the "flux time." The criticality search is performed by regulating the position (in the Z-direction) of the control rods. It is assumed that a control rod consists of the absorber (upper part) and the follower (lower part), and that the response matrices of the control rod are expressed by a linear combination of the response matrices of the upper and lower parts.

Let $\mathbf{R}_u$ and $\mathbf{R}_l$ represent the response matrices of the upper and lower parts, respectively. Then the response matrix of the entire rod is assumed to be expressed by

$$\mathbf{R} = (1 - \gamma)\mathbf{R}_u + \gamma\mathbf{R}_l, \qquad 0 \leq \gamma \leq 1, \tag{8.91}$$

where γ is a parameter representing the position of the control rod. $\gamma = 0$ corresponds to the full insertion of the absorber part into the core, and $\gamma = 1$ corresponds to the full removal of the absorber part and replacement by the follower.

In the process of criticality search, γ is adjusted so that the reactor becomes just critical. When γ becomes unity, the next specified control

* Programmed by the authors.

rod is regulated to attain the criticality. Three types of the removal pattern are programmed in MERMAID-2B.

Besides the value of γ, the mean flux, power density, and flux time in each component rod are obtained as outputs. All the programs and the data are stored simultaneously in the core storage of the IBM-7090.

8.6.1 An Example*

Now let us complete our study by observing the results of a burnup calculation performed for the detailed design of the materials testing reactor referred to in Section 6.1.

In the example given in Table 8.6, the reactor was divided into 20×20 component rods as shown in Fig. 6.1. One "time step" corresponds to 4 days of continuous operation with a thermal output of 30 MW. The control rods A_1 and A_3 were simultaneously removed (represented by γ_1 in the table). After full removal of A_1 and A_3, the control rod A_0 was removed (γ_2). Other control rods were fully removed from the beginning; in other words, they were replaced by the Be follower or by the fuel follower. At each time step the criticality search was performed until $| \lambda - 1 | < 0.005$ was satisfied. At time step 12, γ_2 exceeded unity and interpolation was performed to obtain $\gamma_2 = 1$; thus, time step 13 corresponds to the end of core life (43.05 days).

Table 8.6 also gives the thermal neutron flux at an irradiation hole (3SW in Fig. 6.1), the power density at the center of the core (average power density in the fuel assembly located between A_0 and A_3) and near the reflector (the fuel assembly at the right of A_2).

Computation time for this problem was about 40 minutes. This corresponds to the time required by conventional methods to perform only one or two "criticality calculations."

The reasons for the rapidity of MERMAID-2B are analyzed as follows:

(1) The rapidity of each criticality calculation.

(2) The response matrices of nonburnable rods remain constant throughout core life. Accordingly, it is sufficient to recalculate only the response matrices of the burnable rods at the beginning of each time step.

* A series of burnup calculations was made for the design of JMTR (Japan Materials Testing Reactor) by Mr. S. Miyamoto of Nippon Atomic Industry Group Co., Ltd.

Table 8.6 *Burnup Calculation for a Materials Testing Reactor*

Time step	Days	γ_1	γ_2	λ	C.S.[a] no.	$\bar{\phi}_2$ at 3SW ($\times 10^{14}$)	Power density At center (kW/liter)	Power density Near reflector (kW/liter)
1	0	0.447	0.000	1.0043	6	1.344	211	324
2	4	0.569	0.000	1.0034	3	1.350	218	315
3	8	0.654	0.000	1.0003	1	1.366	224	309
4	12	0.741	0.000	1.0005	3	1.382	230	302
5	16	0.817	0.000	0.9968	2	1.396	236	296
6	20	0.901	0.000	1.0009	3	1.412	242	291
7	24	0.974	0.000	1.0002	3	1.432	248	285
8	28	1.000	0.202	1.0004	8	1.457	263	282
9	32	1.000	0.473	1.0002	2	1.475	283	283
10	36	1.000	0.698	1.0018	2	1.506	302	281
11	40	1.000	0.895	1.0038	2	1.531	322	280
12	44	1.000	1.033[b]	0.9963	1	—	—	—
13	43.05	1.000	1.009	1.0049	1	1.554	334	279

[a] Number of criticality searches required to obtain $|\lambda - 1| < 0.005$.
[b] Indicates the reactor is subcritical; interpolation is performed at the next step.

(3) For each time step, the initial value of γ is extrapolated from the results of the previous calculation. In the example, the required number of criticality searches was only two or three except at the initial stage of control rod removal.

(4) For each criticality calculation, the partial currents obtained from the previous one are available as the initial guess. In the above example, the required number of iterations necessary to obtain λ within an accuracy of 0.001 averaged only 16.

(5) All the programs and data are stored in the core storage of the computer.

APPENDIX A *TIME-DEPENDENT PROBLEM*

For the rigorous treatment of noncritical reactors, it is necessary to consider the time-dependent problems (usually called core kinetics). The response matrix method can also be applied to these problems. In the following discussions we consider the time-dependent response of an infinite homogeneous slab. But the results can be extended directly to other geometries.

A.1 Time-Dependent Response Matrices

Let us inject a pulsed current of neutrons $\mathbf{J}^+(0)\,\delta(t)$ through the left end of an infinite homogeneous slab, where $\delta(t)$ is the Dirac delta function. Let $\mathbf{J}^-(0, t)\,dt$ $[\mathbf{J}^+(1, t)\,dt]$ represent the number and energy spectrum of the reflected (transmitted) neutrons within the time interval dt around t. By virtue of the linearity and the stationarity of the response, we can write

$$\mathbf{J}^-(0, t)\,dt = dt\,\mathbf{R}(t)\mathbf{J}^+(0), \tag{A.1}$$

$$\mathbf{J}^+(1, t)\,dt = dt\,\mathbf{T}(t)\mathbf{J}^+(0). \tag{A.2}$$

Hereafter, $\mathbf{R}(t) = \{R_{nm}(t)\}$ and $\mathbf{T}(t) = \{T_{nm}(t)\}$ will be called the "time-dependent response matrices" so as to distinguish them from the "stationary response matrices" defined in Chapter 7.

Since

$$\int_0^\infty \mathbf{J}^-(0, t)\, dt$$

is equivalent to the stationary current $\mathbf{J}^-(0)$ induced by the stationary incident current $\mathbf{J}^+(0)$, we can derive the following relations between the time-dependent and stationary response matrices:

$$R_{nm} = \int_0^\infty dt\, R_{nm}(t), \tag{A.3}$$

$$T_{nm} = \int_0^\infty dt\, T_{nm}(t). \tag{A.4}$$

All the integrals exist and are finite as long as the slab by itself is subcritical.

In dealing with the initial value problems, it is often convenient to use the technique of the Laplace transformation. We shall denote the Laplace transforms of $\mathbf{R}(t)$ and $\mathbf{T}(t)$ by $\bar{\mathbf{R}}(s)$ and $\bar{\mathbf{T}}(s)$, respectively. Namely,

$$\bar{R}_{nm}(s) = \int_0^\infty e^{-st} R_{nm}(t)\, dt, \tag{A.5}$$

$$\bar{T}_{nm}(s) = \int_0^\infty e^{-st} T_{nm}(t)\, dt. \tag{A.6}$$

Later we shall calculate $\bar{\mathbf{R}}(s)$ and $\bar{\mathbf{T}}(s)$ based on the multigroup diffusion theory.

A.2 Law of Synthesis

We can also derive the "law of synthesis" for $\bar{\mathbf{R}}(s)$ and $\bar{\mathbf{T}}(s)$ by considering the time-dependent response of a composite slab consisting of two slabs A and B (cf. Fig. 7.2).

Consider the time-dependent response currents due to a pulsed incident current $\mathbf{J}^+(0)\, \delta(t)$ injected through the left end of slab A. Using the same notation as in Fig. 7.2, we find following equations with respect to slab A:

$$\mathbf{J}^-(0, t) = \mathbf{R}_A{}^+(t)\mathbf{J}^+(0) + \int_0^t d\tau\, \mathbf{T}_A{}^-(t-\tau)\mathbf{J}^-(1, \tau), \tag{A.7}$$

$$\mathbf{J}^+(1, t) = \mathbf{T}_A{}^+(t)\mathbf{J}^+(0) + \int_0^t d\tau\, \mathbf{R}_A{}^-(t-\tau)\mathbf{J}^-(1, \tau). \tag{A.8}$$

Similarly, for slab B, we have

$$\mathbf{J}^-(1, t) = \int_0^t d\tau\, \mathbf{R}_B{}^+(t-\tau)\mathbf{J}^+(1, \tau), \tag{A.9}$$

$$\mathbf{J}^+(2, t) = \int_0^t d\tau\, \mathbf{T}_B{}^+(t-\tau)\mathbf{J}^+(1, \tau). \tag{A.10}$$

The application of the Laplace transformation will simplify the above equations, because of the "convolution theorem." For instance,*

$$\int_0^\infty e^{-st}\left[\int_0^t \mathbf{T}_A{}^-(t-\tau)\mathbf{J}^-(1, \tau)\, d\tau\right] dt = \bar{\mathbf{T}}_A{}^-(s)\bar{\mathbf{J}}^-(1, s).$$

Thus we get

$$\bar{\mathbf{J}}^-(0, s) = \bar{\mathbf{R}}_A{}^+(s)\mathbf{J}^+(0) + \bar{\mathbf{T}}_A{}^-(s)\bar{\mathbf{J}}^-(1, s), \tag{A.11}$$

$$\bar{\mathbf{J}}^+(1, s) = \bar{\mathbf{T}}_A{}^+(s)\mathbf{J}^+(0) + \bar{\mathbf{R}}_A{}^-(s)\bar{\mathbf{J}}^-(1, s), \tag{A.12}$$

$$\bar{\mathbf{J}}^-(1, s) = \bar{\mathbf{R}}_B{}^+(s)\bar{\mathbf{J}}^+(1, s), \tag{A.13}$$

$$\bar{\mathbf{J}}^+(2, s) = \bar{\mathbf{T}}_B{}^+(s)\bar{\mathbf{J}}^+(1, s), \tag{A.14}$$

and elimination of $\bar{\mathbf{J}}^+(1, s)$ and $\bar{\mathbf{J}}^-(1, s)$ gives

$$\bar{\mathbf{J}}^-(0, s) = \bar{\mathbf{R}}^+_{A+B}(s)\mathbf{J}^+(0), \tag{A.15}$$

$$\bar{\mathbf{J}}^+(2, s) = \bar{\mathbf{T}}^+_{A+B}(s)\mathbf{J}^+(0), \tag{A.16}$$

where

$$\bar{\mathbf{R}}^+_{A+B}(s) = \bar{\mathbf{R}}_A{}^+(s) + \bar{\mathbf{T}}_A{}^-(s)[\mathbf{E} - \bar{\mathbf{R}}_B{}^+(s)\bar{\mathbf{R}}_A{}^-(s)]^{-1}\bar{\mathbf{R}}_B{}^+(s)\bar{\mathbf{T}}_A{}^+(s), \tag{A.17}$$

$$\bar{\mathbf{T}}^+_{A+B}(s) = \bar{\mathbf{T}}_B{}^+(s)[\mathbf{E} - \bar{\mathbf{R}}_A{}^-(s)\bar{\mathbf{R}}_B{}^+(s)]^{-1}\bar{\mathbf{T}}_A{}^+(s). \tag{A.18}$$

* Clearly $\mathbf{T}_A{}^-(t)$ vanishes for negative values of t.

Similarly, $\bar{\mathbf{R}}^-_{A+B}(s)$ and $\bar{\mathbf{T}}^-_{A+B}(s)$ can be written as

$$\bar{\mathbf{R}}^-_{A+B}(s) = \bar{\mathbf{R}}_B^-(s)+\bar{\mathbf{T}}_B^+(s)[\mathbf{E} - \bar{\mathbf{R}}_A^-(s)\bar{\mathbf{R}}_B^+(s)]^{-1}\bar{\mathbf{R}}_A^-(s)\bar{\mathbf{T}}_B^-(s), \tag{A.19}$$

$$\bar{\mathbf{T}}^-_{A+B}(s) = \bar{\mathbf{T}}_A^-(s)[\mathbf{E} - \bar{\mathbf{R}}_B^+(s)\bar{\mathbf{R}}_A^-(s)]^{-1}\bar{\mathbf{T}}_B^-(s). \tag{A.20}$$

Thus we have shown that the Laplace transforms of the time-dependent matrices obey the same "law of synthesis" as that for the stationary response matrices derived in Chapter 7 [cf. Eqs. (7.18)–(7.21)].

A.3 Periods of a Slab

Generally, the equation for the time-dependent flux $\phi(x, t)$ in a slab is linear provided the nuclear characteristics of the slab do not change with time. Thence $\phi(x, t)$, and accordingly $\mathbf{R}(t)$ and $\mathbf{T}(t)$, can each be represented by a linear combination of exponential functions $\{\exp(\omega^{(0)} \cdot t), \exp(\omega^{(1)} \cdot t), \ldots\}$, where $\{\omega^{(0)}, \omega^{(1)}, \ldots\}$ are scalar constants. We shall call $\{\omega^{(k)}\}$ the "inverse periods" and $\{1/\omega^{(k)}\}$ the "periods" of the slab.

In fact, the inverse transformations of $\bar{R}_{nm}(s)$ and $\bar{T}_{nm}(s)$ give

$$R_{nm}(t) = \sum_k \alpha_{nm}^{(k)} \exp(\omega_{nm}^{(k)} \cdot t), \tag{A.21}$$

$$T_{nm}(t) = \sum_k \beta_{nm}^{(k)} \exp(\omega_{nm}^{(k)} \cdot t), \tag{A.22}$$

where $\omega_{nm}^{(k)}$ $(k = 0, 1, 2, \ldots)$ are determined as the poles of $\bar{R}_{nm}(s)$ in the s-plane,* and $\alpha_{nm}^{(k)}$ and $\beta_{nm}^{(k)}$ are determined from the corresponding residues.

If the slab contains fissionable materials, $\{\omega_{nm}^{(k)}\}$ become independent of n and m. Thus, in this case we can write

$$\mathbf{R}(t) = \sum_k \mathbf{A}^{(k)} \exp(\omega^{(k)} \cdot t), \tag{A.23}$$

$$\mathbf{T}(t) = \sum_k \mathbf{B}^{(k)} \exp(\omega^{(k)} \cdot t). \tag{A.24}$$

* Poles of $\bar{T}_{nm}(s)$ coincide with those of $\bar{R}_{nm}(s)$.

A.4 Periods of a Reactor

The inverse periods of a composite slab $A + B$ are determined as the poles of $\bar{\mathbf{R}}^{\pm}_{A+B}(s)$ or $\bar{\mathbf{T}}^{\pm}_{A+B}(s)$. Further, from Eqs. (A.17)–(A.20), we can say that they are determined by

$$| \mathbf{E} - \bar{\mathbf{R}}_A^-(\omega)\bar{\mathbf{R}}_B^+(\omega) | = 0. \tag{A.25}$$

Similarly, the inverse periods of a reactor which can be regarded as a combined system of two subassemblies are given by

$$| \mathbf{E} - \bar{\mathbf{R}}_l^-(\omega)\bar{\mathbf{R}}_r^+(\omega) | = 0. \tag{A.26}$$

From a physical point of view, we require the largest ω to be zero for a just-critical reactor. We already know that $\bar{\mathbf{R}}(0) = \mathbf{R}$ $[\bar{\mathbf{T}}(0) = \mathbf{T}]$ is true for every matrix [cf. Eqs. (A.3)–(A.6)]. Thus it is now proved that (A.26) and (7.25) are equivalent for a critical reactor.

A.5 Rigorous Treatment of a Noncritical Reactor

Let $\mathbf{J}_0^+ \, \delta(t)$ be the pulsed current injected into the right subassembly through the dividing surface (cf. Fig. 7.3). Then it can be shown that the time-dependent currents at the dividing surface are given by

$$\mathbf{J}^+(t) = \mathscr{L}^{-1}\{[\mathbf{E} - \bar{\mathbf{R}}_l^-(s)\bar{\mathbf{R}}_r^+(s)]^{-1} \cdot \mathbf{J}_0^+\}, \tag{A.27}$$

$$\mathbf{J}^-(t) = \mathscr{L}^{-1}\{\bar{\mathbf{R}}_r^+(s) \cdot [\mathbf{E} - \bar{\mathbf{R}}_l^-(s)\bar{\mathbf{R}}_r^+(s)]^{-1} \cdot \mathbf{J}_0^+\}, \tag{A.28}$$

where the pulsed incident current is included in $\mathbf{J}^+(t)$, and $\mathscr{L}^{-1}$ represents the "inverse Laplace transformation."

For sufficiently large t, the above equations are reduced to

$$\mathbf{J}^{\pm}(t) = \exp(\omega^{(0)} \cdot t) \cdot \mathbf{J}^{\pm}, \tag{A.29}$$

where $\omega^{(0)}$ is the largest ω satisfying (A.26). The $\mathbf{J}^{\pm}$ are constant vectors unique to the reactor and are determined by the equations

$$[\mathbf{E} - \bar{\mathbf{R}}_l^-(\omega^{(0)})\bar{\mathbf{R}}_r^+(\omega^{(0)})] \cdot \mathbf{J}^+ = 0 \tag{A.30}$$

and

$$\mathbf{J}^- = \bar{\mathbf{R}}_r^+(\omega^{(0)}) \cdot \mathbf{J}^+. \tag{A.31}$$

Equations (A.30) and (A.31) are the rigorous equations for determining the characteristic spectrum of the currents in a noncritical reactor.

A.6 Approximate Relation between λ and ω

Let us assume that $\mathbf{R}_r^+(t)$ and $\mathbf{R}_l^-(t)$ are expressed by

$$\mathbf{R}_r^+(t) = f_r(t)\mathbf{R}_r^+, \tag{A.32}$$

$$\mathbf{R}_l^-(t) = f_l(t)\mathbf{R}_l^-, \tag{A.33}$$

where $f_r(t)$ and $f_l(t)$ are scalar functions satisfying

$$\int_0^\infty f_r(t)\,dt = 1 \qquad \text{and} \qquad \int_0^\infty f_l(t)\,dt = 1.$$

By transforming (A.32) and (A.33) and substituting the results into (A.26) and (A.30), we obtain the following equations which are identical to (7.24) and (7.25):

$$|\,\lambda\mathbf{E} - \mathbf{R}_l^-\mathbf{R}_r^+\,| = 0 \tag{A.34}$$

and

$$(\lambda\mathbf{E} - \mathbf{R}_l^-\mathbf{R}_r^+) \cdot \mathbf{J}^+ = 0 \tag{A.35}$$

with

$$\lambda = \frac{1}{\bar{f}_l(\omega) \cdot \bar{f}_r(\omega)}. \tag{A.36}$$

For a special selection,

$$f_r(t) = -\omega_r \exp(\omega_r \cdot t) \qquad (\omega_r < 0),$$

$$f_l(t) = -\omega_l \exp(\omega_l \cdot t) \qquad (\omega_l < 0),$$

Eq. (A.36) becomes

$$\lambda = \frac{(\omega_r - \omega)(\omega_l - \omega)}{\omega_r\omega_l}, \tag{A.37}$$

where ω_r and ω_l are, respectively, the largest inverse period of the right and left subassemblies. The largest inverse period of the reactor is then given by

$$\omega^{(0)} = \tfrac{1}{2}\{(\omega_l + \omega_r) + [(\omega_l + \omega_r)^2 + 4\omega_l\omega_r(\lambda - 1)]^{1/2}\}. \quad \text{(A.38)}$$

Particularly, if $\omega_l = \omega_r$, we have

$$\omega^{(0)} = (1 - \lambda^{1/2})\omega_l. \quad \text{(A.39)}$$

A.7 Calculation of $\bar{\mathbf{R}}(s)$ and $\bar{\mathbf{T}}(s)$

Taking the effect of delayed neutrons into account, we shall calculate $\bar{\mathbf{R}}(s)$ and $\bar{\mathbf{T}}(s)$ for a homogeneous slab of thickness a based on G-group diffusion theory. Let $\phi_n(x, t)$ be the neutron flux of group n due to the pulsed current of group m $[J_m^+(0)\,\delta(t)]$ injected through the surface $x = 0$. Then $\phi_n(x, t)$ satisfy the time-dependent diffusion equation

$$\begin{aligned}
\frac{1}{v_1}\frac{\partial\phi_1}{\partial t} &= D_1\frac{\partial^2}{\partial x^2}\phi_1 - \Sigma_1\phi_1 + (1-\beta)\nu\Sigma_{fG}\phi_G + \sum_{i=1}^{I}\lambda_i C_i(x, t),\\
\frac{1}{v_2}\frac{\partial\phi_2}{\partial t} &= D_2\frac{\partial^2}{\partial x^2}\phi_2 - \Sigma_2\phi_2 + \Sigma_{s1}\phi_1,\\
&\;\;\vdots\\
\frac{1}{v_G}\frac{\partial\phi_G}{\partial t} &= D_G\frac{\partial^2}{\partial x^2}\phi_G - \Sigma_G\phi_G + \Sigma_{s\,G-1}\phi_{G-1},
\end{aligned} \quad \text{(A.40)}$$

$$\frac{\partial C_i}{\partial t} = -\lambda_i C_i(x, t) + \beta_i\nu\Sigma_{fG}\phi_G \qquad (i = 1\text{–}I), \quad \text{(A.41)}$$

with the boundary conditions

$$\begin{aligned}
J_n^-(a, t) &= \frac{1}{4}\left\{\phi_n(x, t) + d_n\frac{\partial}{\partial x}\phi_n(x, t)\right\}_{x=a} = 0 \qquad (n = 1\text{–}G),\\
J_n^+(0, t) &= \frac{1}{4}\left\{\phi_n(x, t) - d_n\frac{\partial}{\partial x}\phi_n(x, t)\right\}_{x=0} = 0 \qquad (n \neq m),\\
J_m^+(0, t) &= \frac{1}{4}\left\{\phi_m(x, t) - d_m\frac{\partial}{\partial x}\phi_m(x, t)\right\}_{x=0} = J_m^+(0)\,\delta(t),
\end{aligned} \quad \text{(A.42)}$$

and the initial conditions

$$\phi_n(x, 0) = 0, \qquad C_i(x, 0) = 0, \tag{A.43}$$

where I groups of "delayed neutron precursors" are considered, and it is assumed that the velocity of all the delayed neutrons is equal to that of the energy group 1. The following notations are used in the above and the following equations:

v_g	average velocity of neutrons of energy group g,
$C_i(x, t)$	concentration of the precursors of the ith group,
β_i	probability for the generation of precursors belonging to the ith group by a fission,
β	$= \sum_{i=1}^{I} \beta_i$,
λ_i	decay constant for the ith group.

Performing the Laplace transformations of Eqs. (A.40), (A.41), and (A.42), we obtain the following equations and boundary conditions:

$$\begin{aligned} -D_1 \frac{\partial^2}{\partial x^2} \bar{\phi}_1 + \left(\Sigma_1 + \frac{s}{v_1}\right)\bar{\phi}_1 &= \nu\Sigma_{fG}\left\{1 - \beta + \sum_{i=1}^{I} \frac{\lambda_i \beta_i}{\lambda_i + s}\right\}\bar{\phi}_G, \\ -D_2 \frac{\partial^2}{\partial x^2} \bar{\phi}_2 + \left(\Sigma_2 + \frac{s}{v_2}\right)\bar{\phi}_2 &= \Sigma_{s1}\bar{\phi}_1, \\ &\vdots \\ -D_G \frac{\partial^2}{\partial x^2} \bar{\phi}_G + \left(\Sigma_G + \frac{s}{v_G}\right)\bar{\phi}_G &= \Sigma_{s\,G-1}\bar{\phi}_{G-1}, \end{aligned} \tag{A.44}$$

$$\begin{aligned} \frac{1}{4}\left\{\bar{\phi}_n(x, s) + d_n \frac{\partial}{\partial x} \bar{\phi}_n(x, s)\right\}_{x=a} &= 0, \\ \frac{1}{4}\left\{\bar{\phi}_n(x, s) - d_n \frac{\partial}{\partial x} \bar{\phi}_n(x, s)\right\}_{x=0} &= \delta_{nm} \cdot J_m^{+}(0), \end{aligned} \tag{A.45}$$

where

$$\bar{\phi}_n(x, s) = \int_0^\infty dt\, e^{-st}\phi_n(x, t),$$

and we have used the relation

$$\bar{C}_i(x, s) = \frac{\beta_i \nu \Sigma_{fG} \bar{\phi}_G(x, s)}{\lambda_i + s}.$$

It can easily be observed that the above equations and boundary conditions for $\bar{\phi}_n(x, s)$ are identical to those for the time-independent flux given in Chapter 7, except that Σ_g and $\nu\Sigma_{fG}$ are now replaced by

$$\left(\Sigma_g + \frac{s}{v_g}\right) \quad \text{and} \quad \nu\Sigma_{fG}\left\{1 - \beta + \sum_{i=1}^{I} \frac{\lambda_i\beta_i}{\lambda_i + s}\right\},$$

respectively [cf. Eqs. (7.51), (7.56), and (7.57)].

Since $\bar{R}_{nm}(s)$ and $\bar{T}_{nm}(s)$ are given by

$$\bar{R}_{nm}(s) = \frac{\bar{J}^-_{nm}(0, s)}{J_m^+(0)} = \frac{\{\bar{\phi}_n(x, s) + d_n(\partial/\partial x)\bar{\phi}_n(x, s)\}_{x=0}}{4J_m^+(0)}, \tag{A.46}$$

$$\bar{T}_{nm}(s) = \frac{\bar{J}^+_{nm}(a, s)}{J_m^+(0)} = \frac{\{\bar{\phi}_n(x, s) - d_n(\partial/\partial x)\bar{\phi}_n(x, s)\}_{x=a}}{4J_m^+(0)}, \tag{A.47}$$

we can conclude that the matrices $\bar{\mathbf{R}}(s)$ and $\bar{\mathbf{T}}(s)$ are readily calculated from the formulas for the stationary matrices $\mathbf{R}$ and $\mathbf{T}$ with the abovementioned replacements. For example, $\bar{R}(s)$ and $\bar{T}(s)$ with one-group diffusion theory are derived from Eqs. (7.40) and (7.41),

$$\bar{R}(s) = \frac{(1 + K^2d^2)\sin Ka}{(1 - K^2d^2)\sin Ka + 2Kd\cos Ka}, \tag{A.48}$$

$$\bar{T}(s) = \frac{2Kd}{(1 - K^2d^2)\sin Ka + 2Kd\cos Ka}, \tag{A.49}$$

where

$$K = K(s) = \left(\frac{1}{D}\left[\left\{(1 - \beta) + \sum_{i=1}^{I} \frac{\lambda_i\beta_i}{\lambda_i + s}\right\}\nu\Sigma_f - \left(\Sigma_a + \frac{s}{v}\right)\right]\right)^{1/2}. \tag{A.50}$$

APPENDIX B $\tilde{\mathbf{R}}$ *AND* $\tilde{\mathbf{T}}$ *USING TWO-GROUP DIFFUSION APPROXIMATION*

The formulas for calculating the elements of $\tilde{\mathbf{R}}$ and $\tilde{\mathbf{T}}$ for a homogeneous slab of thickness a using the two-group diffusion approximation follow. The exact response matrices **R** and **T** are easily obtained from Eqs. (7.59) and (7.60).

B.1 The Case $k_\infty \geq 1$

$$
\begin{aligned}
\tilde{R}_{11} &= \frac{1}{\Delta}\,[S_\mu(1 + \mu d_2 \cot \mu a)(1 - \nu d_1 \coth \nu a) \\
&\qquad + S_\nu(1 - \mu d_1 \cot \mu a)(1 + \nu d_2 \coth \nu a)], \\
\tilde{R}_{12} &= \frac{2}{\Delta}\,[\nu d_1 \coth \nu a - \mu d_1 \cot \mu a], \\
\tilde{R}_{21} &= \frac{2 S_\mu S_\nu}{\Delta}\,[\nu d_2 \coth \nu a - \mu d_2 \cot \mu a], \\
\tilde{R}_{22} &= \frac{1}{\Delta}\,[S_\mu(1 - \mu d_2 \cot \mu a)(1 + \nu d_1 \coth \nu a) \\
&\qquad + S_\nu(1 + \mu d_1 \cot \mu a)(1 - \nu d_2 \coth \nu a)];
\end{aligned}
\tag{B.1}
$$

$$\tilde{T}_{11} = \frac{1}{\Delta}\left[\frac{\nu d_1 S_\mu}{\sinh \nu a}(1 + \mu d_2 \cot \mu a) + \frac{\mu d_1 S_\nu}{\sin \mu a}(1 + \nu d_2 \coth \nu a)\right],$$

$$\tilde{T}_{12} = \frac{1}{\Delta}\left[\frac{\mu d_1}{\sin \mu a}(1 + \nu d_1 \coth \nu a) - \frac{\nu d_1}{\sinh \nu a}(1 + \mu d_1 \cot \mu a)\right],$$

$$\tilde{T}_{21} = \frac{S_\mu S_\nu}{\Delta}\left[\frac{\mu d_2}{\sin \mu a}(1+\nu d_2 \coth \nu a) - \frac{\nu d_2}{\sinh \nu a}(1+\mu d_2 \cot \mu a)\right], \tag{B.2}$$

$$\tilde{T}_{22} = \frac{1}{\Delta}\left[\frac{\mu d_2 S_\mu}{\sin \mu a}(1 + \nu d_1 \coth \nu a) + \frac{\nu d_2 S_\nu}{\sinh \nu a}(1 + \mu d_1 \cot \mu a)\right];$$

where

$$d_1 = 2D_1, \qquad d_2 = 2D_2,$$

$$\Delta = S_\mu(1 + \mu d_2 \cot \mu a)(1 + \nu d_1 \coth \nu a) + S_\nu(1 + \mu d_1 \cot \mu a)(1 + \nu d_2 \coth \nu a);$$

μ and ν are, respectively, the positive roots of the equations

$$\mu^2 = \frac{1}{2}\left\{-\left(\frac{1}{L_1^2} + \frac{1}{L_2^2}\right) + \left[\left(\frac{1}{L_1^2} + \frac{1}{L_2^2}\right)^2 + \frac{4(k_\infty - 1)}{L_1^2 L_2^2}\right]^{1/2}\right\} \tag{B.3}$$

and

$$\nu^2 = \frac{1}{2}\left\{\left(\frac{1}{L_1^2} + \frac{1}{L_2^2}\right) + \left[\left(\frac{1}{L_1^2} + \frac{1}{L_2^2}\right)^2 + \frac{4(k_\infty - 1)}{L_1^2 L_2^2}\right]^{1/2}\right\}; \tag{B.4}$$

$$S_\mu = \left(\frac{D_1}{D_2}\right) P \frac{1}{L_1^2} \frac{1}{\mu^2 + (1/L_2^2)},$$

$$S_\nu = \left(\frac{D_1}{D_2}\right) P \frac{1}{L_1^2} \frac{1}{\mu^2 + (1/L_1^2)},$$

$$P = \frac{\Sigma_{1s}}{\Sigma_1}.$$

B.2 The Case $0 < k_\infty < 1$

The following results are obtained from the above equations by replacing μ by $i\mu$. The equations are

$$\begin{aligned}
\tilde{R}_{11} &= \frac{1}{\Delta}\,[S_\mu(1 + \mu d_2 \coth \mu a)(1 - \nu d_1 \coth \nu a) \\
&\qquad + S_\nu(1 - \mu d_1 \coth \mu a)(1 + \nu d_2 \coth \nu a)], \\
\tilde{R}_{12} &= \frac{2}{\Delta}\,[\nu d_1 \coth \nu a - \mu d_1 \coth \mu a], \\
\tilde{R}_{21} &= \frac{2S_\mu S_\nu}{\Delta}\,[\nu d_2 \coth \nu a - \mu d_2 \coth \mu a], \\
\tilde{R}_{22} &= \frac{1}{\Delta}\,[S_\mu(1 - \mu d_2 \coth \mu a)(1 + \nu d_1 \coth \nu a) \\
&\qquad + S_\nu(1 + \mu d_1 \coth \mu a)(1 - \nu d_2 \coth \nu a)];
\end{aligned} \tag{B.5}$$

$$\begin{aligned}
\tilde{T}_{11} &= \frac{1}{\Delta}\left[\frac{\nu d_1 S_\mu}{\sinh \nu a}(1 + \mu d_2 \coth \mu a) + \frac{\mu d_1 S_\nu}{\sinh \mu a}(1 + \nu d_2 \coth \nu a)\right], \\
\tilde{T}_{12} &= \frac{1}{\Delta}\left[\frac{\mu d_1}{\sinh \mu a}(1+\nu d_1 \coth \nu a) - \frac{\nu d_1}{\sinh \nu a}(1+\mu d_1 \coth \mu a)\right], \\
\tilde{T}_{21} &= \frac{S_\mu S_\nu}{\Delta}\left[\frac{\mu d_2}{\sinh \mu a}(1+\nu d_2 \coth \nu a) - \frac{\nu d_2}{\sinh \nu a}(1+\mu d_2 \coth \mu a)\right], \\
\tilde{T}_{22} &= \frac{1}{\Delta}\left[\frac{\mu d_2 S_\mu}{\sinh \mu a}(1 + \nu d_1 \coth \nu a) + \frac{\nu d_2 S_\nu}{\sinh \nu a}(1 + \mu d_1 \coth \mu a)\right];
\end{aligned} \tag{B.6}$$

where

$$\begin{aligned}
\Delta &= S_\mu(1 + \mu d_2 \coth \mu a)(1 + \nu d_1 \coth \nu a) \\
&\qquad + S_\nu(1 + \mu d_1 \coth \mu a)(1 + \nu d_2 \coth \nu a),
\end{aligned}$$

$$\mu^2 = \frac{1}{2}\left\{\left(\frac{1}{L_1^2} + \frac{1}{L_2^2}\right)^2 - \left[\left(\frac{1}{L_1^2} + \frac{1}{L_2^2}\right)^2 + \frac{4(k_\infty - 1)}{L_1^2 L_2^2}\right]^{1/2}\right\},$$

$$\nu^2 = \frac{1}{2}\left\{\left(\frac{1}{L_1^2} + \frac{1}{L_2^2}\right)^2 + \left[\left(\frac{1}{L_1^2} + \frac{1}{L_2^2}\right)^2 + \frac{4(k_\infty - 1)}{L_1^2 L_2^2}\right]^{1/2}\right\},$$

$$S_\mu = \left(\frac{D_1}{D_2}\right) P \frac{1}{L_1^2}\,\frac{1}{(1/L_2^2) - \mu^2},$$

$$S_\nu = \left(\frac{D_1}{D_2}\right) P \frac{1}{L_1^2}\,\frac{1}{(1/L_1^2) - \mu^2}.$$

B.3 The Case $k_\infty = 0$

$$\tilde{R}_{11} = \frac{1 - \dfrac{d_1}{L_1}\coth\dfrac{a}{L_1}}{1 + \dfrac{d_1}{L_1}\coth\dfrac{a}{L_1}},$$

$$\tilde{R}_{12} = 0,$$

$$\tilde{R}_{21} = \frac{2S\left(\dfrac{d_2}{L_2}\coth\dfrac{a}{L_2} - \dfrac{d_1}{L_1}\coth\dfrac{a}{L_1}\right)}{\left(1 + \dfrac{d_1}{L_1}\coth\dfrac{a}{L_1}\right)\left(1 + \dfrac{d_2}{L_2}\coth\dfrac{a}{L_2}\right)}, \tag{B.7}$$

$$\tilde{R}_{22} = \frac{1 - \dfrac{d_2}{L_2}\coth\dfrac{a}{L_2}}{1 + \dfrac{d_2}{L_2}\coth\dfrac{a}{L_2}};$$

$$\tilde{T}_{11} = \frac{\dfrac{d_1}{L_1}}{\sinh\dfrac{a}{L_1} + \dfrac{d_1}{L_1}\cosh\dfrac{a}{L_1}},$$

$$\tilde{T}_{12} = 0,$$

$$\tilde{T}_{21} = S\left[\frac{\dfrac{d_2}{L_1}}{\sinh\dfrac{a}{L_1} + \dfrac{d_1}{L_1}\cosh\dfrac{a}{L_1}} - \frac{\dfrac{d_2}{L_2}\left(1 + \dfrac{d_2}{L_1}\coth\dfrac{a}{L_1}\right)}{\left(\sinh\dfrac{a}{L_2} + \dfrac{d_2}{L_2}\cosh\dfrac{a}{L_2}\right)\left(1 + \dfrac{d_1}{L_1}\coth\dfrac{a}{L_1}\right)}\right], \tag{B.8}$$

$$\tilde{T}_{22} = \frac{\dfrac{d_2}{L_2}}{\sinh\dfrac{a}{L_2} + \dfrac{d_2}{L_2}\cosh\dfrac{a}{L_2}},$$

where

$$S = \left(\frac{D_1}{D_2}\right) P \frac{1}{L_1^2} \frac{1}{\dfrac{1}{L_2^2} - \dfrac{1}{L_1^2}}.$$

For a very thick slab ($a \gg L_1, L_2$), the following formulas are useful:

$$\tilde{R}_{11} \simeq \frac{1 - \dfrac{d_1}{L_1}}{1 + \dfrac{d_1}{L_1}},$$

$$\tilde{R}_{21} \simeq \frac{2P\left(\dfrac{d_2}{L_1}\right)}{\left(1 + \dfrac{d_1}{L_1}\right)\left(1 + \dfrac{d_2}{L_2}\right)\left(\dfrac{L_1}{L_2} + 1\right)}, \tag{B.9}$$

$$\tilde{R}_{22} \simeq \frac{1 - \dfrac{d_2}{L_2}}{1 + \dfrac{d_2}{L_2}}.$$

REFERENCES

1. A. M. Weinberg and E. P. Wigner, "The Physical Theory of Neutron Chain Reactors." Univ. of Chicago Press, Chicago, 1958.
2. S. Glasstone and M. C. Edlund, "The Elements of Nuclear Reactor Theory." Van Nostrand-Reinhold, Princeton, New Jersey, 1952.
3. G. E. Forsythe, and W. R. Wasow, "Finite-Difference Methods for Partial Differential Equations." Wiley, New York, 1960.
4. E. L. Wachspress, "Iterative Solution of Elliptic Systems." Prentice-Hall, Englewood Cliffs, New Jersey, 1966.
5. M. Clark and K. F. Hansen, "Numerical Methods of Reactor Analysis." Academic Press, New York, 1964.
6. W. C. Sangren, "Digital Computers and Nuclear Reactor Calculations." Wiley, New York, 1960.
7. G. G. Bilodeau, W. R. Cadwell, J. P. Dorsey, J. M. Fairey, and R. S. Varga, PDQ—An IBM-704 code to solve the two-dimensional few-group neutron-diffusion equations, Rep. WAPD-TM-70 (1957).
8. E. L. Wachspress, CURE: A generalized two-space-dimension multigroup coding for the IBM-704, Rep. KAPL 1724 (1957).
9. R. S. Varga, "Matrix Iterative Analysis." Prentice-Hall, Englewood Cliffs, New Jersey, 1962.
10. S. Kaplan, Some new methods of flux synthesis, *Nuc. Sci. Eng.* **13**, 22 (1962).
11. M. Becker, "The Principles and Applications of Variational Methods." MIT Press, Cambridge, Massachusetts, 1964.
12. D. S. Selengut, Variational analysis of multi-dimensional systems, AEC Rep. HW-59126 (1959).
13. E. Amaldi and E. Fermi, *Ric. Sci.* **7**, 454 (1963).
14. S. G. Stokes, "Mathematical and Physical Papers of Sir George Stokes," Vol. IV, p. 145. Cambridge Univ. Press, London and New York, 1904.
15. R. Bellman, R. Kalaba, and G. M. Wing, On the principle of invariant imbedding and neutron transport theory, I. One-dimensional case, *J. Math. Mech.* **7**, 149 (1958).

16. R. Bellman, R. Kalaba, and G. M. Wing, Invariant imbedding and neutron transport theory, II. Functional equations, *J. Math. Mech.* **7**, 741 (1958).
17. R. Bellman, R. Kalaba, and G. M. Wing, Invariant imbedding and neutron transport theory, III. Neutron–neutron collision processes, *J. Math. Mech.* **8**, 249 (1959).
18. R. Bellman, R. Kalaba, and G. M. Wing, Invariant imbedding and neutron transport theory, IV. Generalized transport theory, *J. Math. Mech.* **8**, 575 (1959).
19. R. Bellman, R. Kalaba, and G. M. Wing, Invariant imbedding and neutron transport theory, V. Diffusion as a limiting case, *J. Math. Mech.* **9**, 933 (1960).
20. R. Bellman, R. Kalaba, and G. M. Wing, Invariant imbedding and neutron transport in a rod of changing length, *Proc. Nat. Acad. Sci. USA* **46**, 128 (1960).
21. R. Bellman, R. Kalaba, and G. M. Wing, Invariant imbedding and mathematical physics, I. Particle processes, *J. Math. Phys.* (*N.Y.*) **1**, 280 (1960).
22. R. Bellman, H. H. Kagiwada, R. E. Kalaba, and M. C. Prestrud, "Invariant Imbedding and Time-Dependent Transport Processes." American Elsevier, New York, 1964.
23. R. T. Ackroyd and J. D. McCullen, Albedo methods, *Proc. Int. Conf. Peaceful Uses At. Energy* **12**, 38 (1958).
24. M. Otsuka, Trapping of Thermal Neutrons, II. Miscellanea, *J. At. Energy Soc. Japan* **4**, 505 (1962).
25. D. S. Selengut, Partial current representations in reactor physics, KAPL-2229 (1963).
26. A. R. Bobrowsky, Analytical method of determining transmission of particles and radiation through great thicknesses of matter, NACA-TN-1712 (1948).
27. A. Shimizu, Response matrix method, *J. At. Energy Soc. Japan* **5**, 359 (1963).
28. A. Shimizu, K. Monta, and T. Miyamoto, Application of response matrix method to criticality calculations of one-dimensional reactors, *J. At. Energy Soc. Japan* **5**, 369 (1963).
29. K. Aoki and A. Shimizu, Application of response matrix method to criticality calculations of two-dimensional reactors, *J. Nucl. Sci. Technol.* **2**, 149 (1965).
30. P. Schmid, Matrix formulation of the two-group diffusion theory for a multiple reflector spherical reactor, *Proc. Int. Conf. Peaceful Uses At. Energy* **5**, 444 (1955).
31. H. Koskinen, Generalized potential theory for multigroup diffusion in a general multiregion reactor, *Proc. Int. Conf. Peaceful Uses At. Energy* **4**, 67 (1965).
32. T. Auerbach, Some applications of Chandrasekhar's method to reactor theory, BNL 676 (1961).
33. S. Chandrasekhar, "Radiative Transfer." Oxford Univ. Press, London and New York, 1950.
34. E. L. Wachspress, Thin regions and diffusion theory calculations, *Nucl. Sci. Eng.* **3**, 186 (1958).
35. H. S. Wilf, The transmission of neutrons in multi-layered slab geometry, *Nucl. Sci. Eng.* **5**, 306 (1956).

AUTHOR INDEX

Numbers in parentheses are reference numbers and indicate that an author's work is referred to, although his name is not cited in the text. Numbers in italics show the page on which the complete reference is listed.

SUBJECT INDEX